A LAND IN MOTION

CALIFORNIA'S SAN ANDREAS FAULT

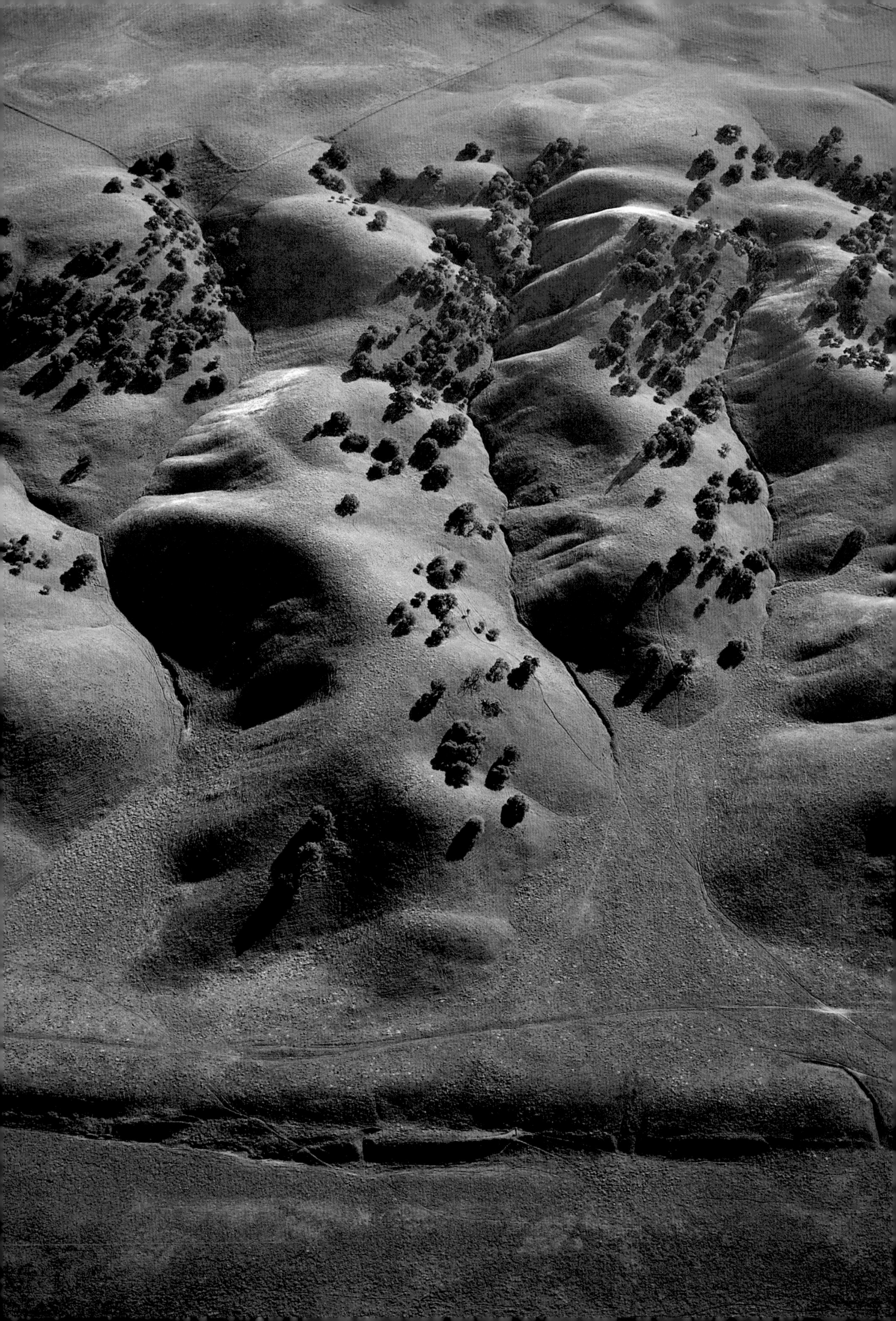

A LAND IN MOTION
CALIFORNIA'S SAN ANDREAS FAULT

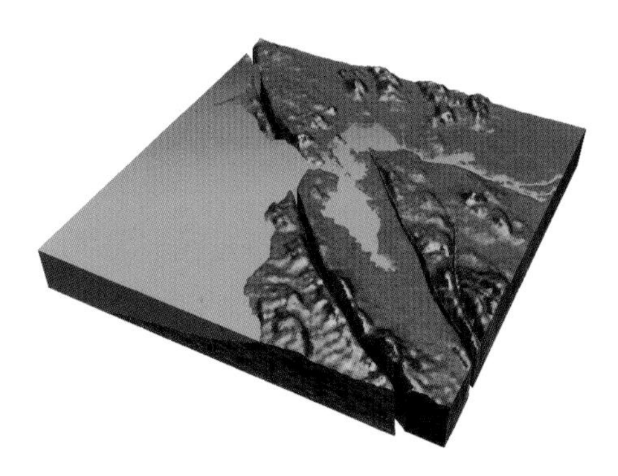

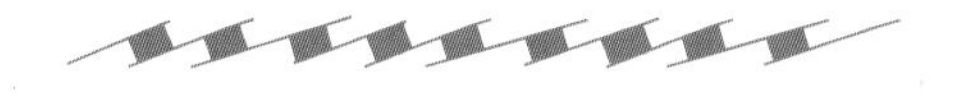

Text and photographs by Michael Collier

Illustrations by Lawrence Ormsby

Golden Gate National Parks Association
San Francisco, California

University of California Press
Berkeley Los Angeles London

in cooperation with the United States Geological Survey
Menlo Park, California

Golden Gate National Parks Association
San Francisco, California

University of California Press
Berkeley and Los Angeles, California

University of California Press, Ltd.
London, England

Library of Congress Catalog Card Number 98-73364
ISBN 0-520-21897-3

Lyrics from "Talkin' to the Rocks," ©Don Charles/Buzzards Luck Music/ASCAP; used with permission
Historical photos (pages 50-51) from the USGS Photographic Library
Earthquake map (page 70) compiled by Stephen Walter, USGS
Landsat photographs (pages 76, 81, and 90-91) compiled by Michael Rymer, USGS
Diagrams (page 110) compiled by Ross Stein, USGS

In creating the illustrations in this book, the artist relied on USGS Professional Paper 1575, "The San Andreas Fault System," by Robert Wallace, with the following exceptions: page 18, after "Essentials of Geology," fourth edition, Lutgens & Tarbuck; page 33, after "The Origin of Allochthonous Terranes: Perspectives on the Growth and Shaping of Continents," Elizabeth Schermer, David Howell, and David E. Jones, *Ann. Rev. Earth Planet. Sci.* 1984, 12:107-31; page 38, after "The Loma Prieta, California, Earthquake...Tectonic Process and Models," USGS Professional Paper 1550-F; page 96, "Is the San Andreas Big Bend responsible for the Landers Earthquake and the eastern California shear zone?" *Geology*, vol. 24, no. 3, pp. 219-22, March 1996; and page 100, after "Measuring Crustal Deformation in the American West," *Scientific American*, August 1988. The illustrator would also like to thank Chris "Xenon" Hanson of 3D Nature for his technical assistance in creating the maps.

Text and illustrations reviewed by United States Geological Survey, Menlo Park, California

Editor: Susan Tasaki
Design, Production: Carole Thickstun

Printed on recycled paper by Precision Litho, Salt Lake City, Utah
9 8 7 6 5 4 3 2 1

Cover: *San Andreas Fault and offset streams on the Carrizo Plain*
Frontis: *Foothills above the San Andreas Fault near Parkfield*
Facing page: *California coastline north of Fort Bragg*
Following spread: *Clouds spilling over the Transverse Ranges south of Santa Inez*

Sharing a mission to promote public information and education, the United States Geological Survey and the National Park Service also have a mutual interest in the San Andreas Fault, which has a direct and dramatic influence on land protected in the five national parks along its path. In response to a call for proposals extended by the USGS, Joe Zarki, Chief of Interpretation at Joshua Tree National Park, suggested the topic that ultimately resulted in this publication. The publishers would like to acknowledge both the National Park Service and the United States Geological Survey's central role in the existence of this book.

If I could move real slow
I could hear the rocks talkin'
If I could move real slow
I could see the trees walkin'
If I could move really slow
I'd listen to the dead talkin'
talkin' to the rocks
walkin' with the trees
comparin' family histories
walkin' with the rocks

"Talkin' to the Rocks"
Don Charles

Contents

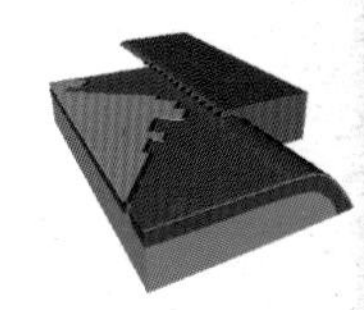

Folds west of the Salton Sea

Fault Lines

I HAD THOUGHT THAT the San Andreas Fault would be more obvious. After all, it is a 750-mile slash that runs the length of California, from near Mexico to the Mendocino coast. The San Andreas is the grinding, growling interface between two great pieces of the earth's crust, each moving its separate way. This is the fault that has cleaved volcanoes, opened seaways, and split mountains. And yet, driving south of Paicines on California Highway 25, I wasn't sure just where it was.

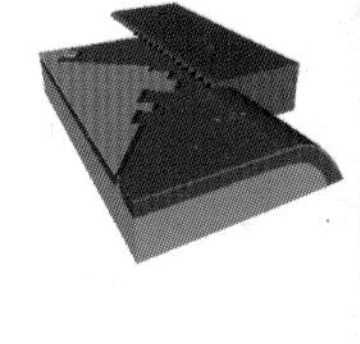

The countryside that rolls alongside the San Benito River is dotted with magnificent oaks, carpeted in lush green or warm brown grass depending on the season. I knew that this river's valley is aligned northwest to southeast, roughly parallel to the San Andreas. The day before, I had flown along the course of the fault from San Bernardino to San Jose. Aloft, the San Andreas stood out boldly, a straight gash across an otherwise tumultuous landscape. Here and there the fault was obscured for a few miles as it snuck across a featureless plain or through tumbling talus. From the cockpit I could look a few miles up or down its course and quickly find it again. But

Farming along the San Andreas Fault, northwest of San Benito

Offset roadbed on California Highway 25 near Paicines

the story was different down here on the ground. Where was that fault?

The road wove around drainages aligned in many directions, not necessarily northwest/southeast. Hillsides terminated with tempting suddenness, but still no solid evidence of the San Andreas. Of course, the rocks exposed in roadcuts changed without warning from one formation to another. In a more coherent geologic province, these changes might have heralded a major fault. But this is California and I had grown accustomed to the unexpected comings and goings of rock layers. Where was the San Andreas?

Thump-thump.

What was that? The car continued to handle steadily, neither smoking nor swaying. I looked in the rearview mirror: no transmission parts were bouncing toward the shoulder. I did see a black hump running diagonally across the road, northwest to southeast. I pulled over and walked back. The highway had been patched repeatedly with asphalt bandaids but the hump persisted. Beyond, the road was offset to the right. Finally I had arrived: the San Andreas.

THE HOPI INDIANS of northern Arizona believe that humans emerged into the Fourth World (the one we now inhabit), through an opening in the earth—a sipapu—in the Grand Canyon. Years ago, I peered down into the sipapu and saw prayer feathers on the wall and carbon dioxide bubbling up through the blue-gray water of a spring. Twist and turn and try as I may, though, I couldn't see the Third World down below. But looking down on the San Andreas, I realize that it *is* possible to see through the geologic looking glass into another world, one of roving continents and subducting plates.

The San Andreas Fault is the product of a collision between the plates that carry the floor of the Pacific Ocean and the North American continent, a collision that began when dinosaurs roamed the earth. This collision was responsible for creation of Sierra Nevada granite and raising the Rocky Mountains; later, it caused the collapse of Death Valley. More recently, the collision ignited Mount Saint Helens and ripped coastal California into geologic confetti. The San Andreas, stretching from the Salton Sea to Point Delgada, is the predominant surface along which the Pacific Plate now moves northwest as the rest of the continent moves roughly west on the North American Plate.

Defined most simply, the San Andreas is a single vertical *fault*—or break—in the earth's crust that extends from the surface to a depth of about ten miles, where rocks are hot enough to ooze rather than fracture. The distinguishing feature that separates a simple fracture from a true fault is movement of rocks on one side relative to the other side. But the San Andreas is not simple. In many places it is more precisely referred to as a *fault zone*, where movement is taken up across a series of parallel (or *en-echelon*) fault segments that may spread out over a mile or more. To further complicate matters, the San Andreas is only one of many faults in California that accommodate movement between the Pacific and North American plates. The other faults, such as the Emerson at Landers, the Hayward

The San Andreas can be elusive from the ground, but it's hard to miss from the air.

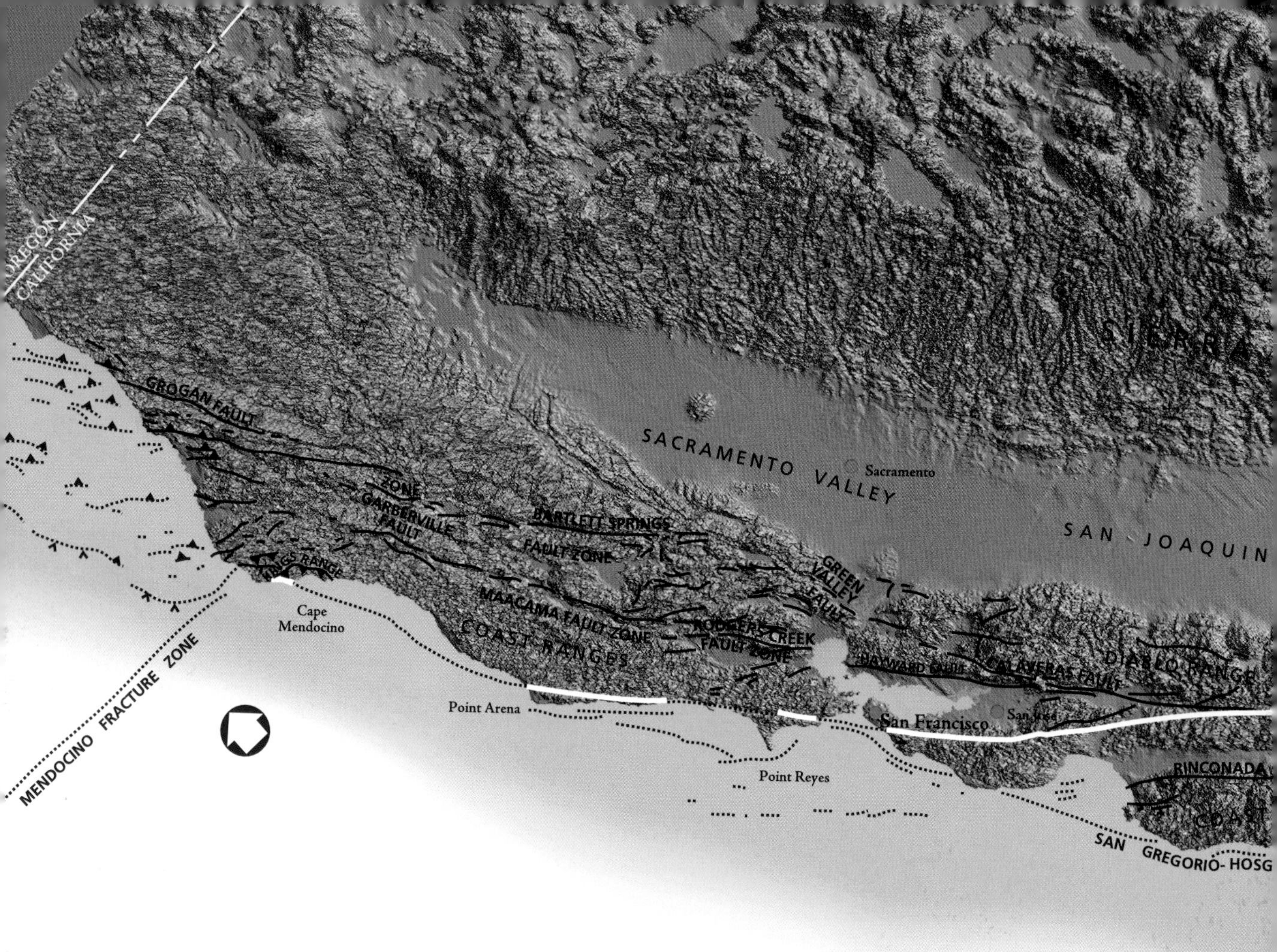

beneath Oakland, and the Elsinore near Anaheim, may not have the length or displacement of the San Andreas, but they are active and they do add to the overall northwest transport of coastal California. Taken together, this collection of faults is inclusively called the San Andreas Fault *system.*

The San Andreas Fault runs beneath Lower Crystal Springs Reservoir and San Andreas Lake alongside Interstate 280.

Andrew Lawson was a young geologist when he coined the phrase "San Andreas Fault." Like all geologists, he knew that straight lines are the exception rather than the rule on the face of the earth. In 1895 he was examining the long narrow San Andreas Valley just south of San Francisco. Too long, too narrow, too straight. Lawson concluded that a fault must run ten or fifteen miles here along the valley sides. Then as now, a dam plugged the valley's downstream end, creating a reservoir that received

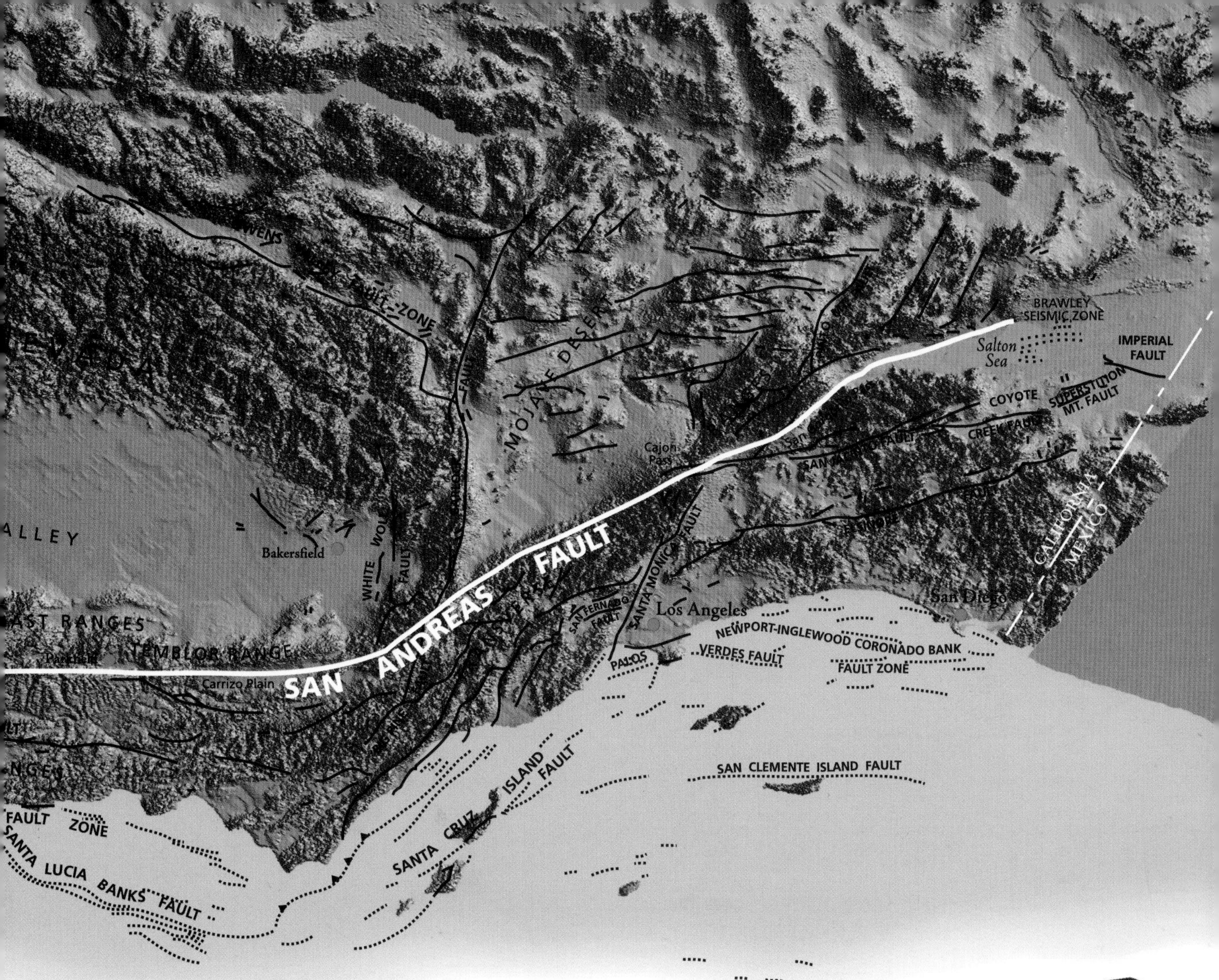

water piped across California's great Central Valley from the Sierra Nevada. Along with the adjacent Crystal Springs Reservoir, this lake remains an integral part of San Francisco's water supply.

Eleven years later, Lawson had occasion to expand his description of the San Andreas. San Francisco was devastated when the fault slipped during the terrible earthquake of 1906. Much that wasn't initially knocked down subsequently burned down. For his 1908 report, published by the Carnegie Institution, Lawson and some of the country's other top geologists tromped not only the 270 miles of northern California that broke in 1906, but also most of what we know today as the full San Andreas Fault. They realized that enlongate coastal mountains frequently paralleled the fault. *It would seem,* he wrote, *that there is sufficient unity of character in these coastal mountains, in spite of their change of trend, to warrant their being classed as the Coast System from South Fork Mountain to the Mexican boundary and*

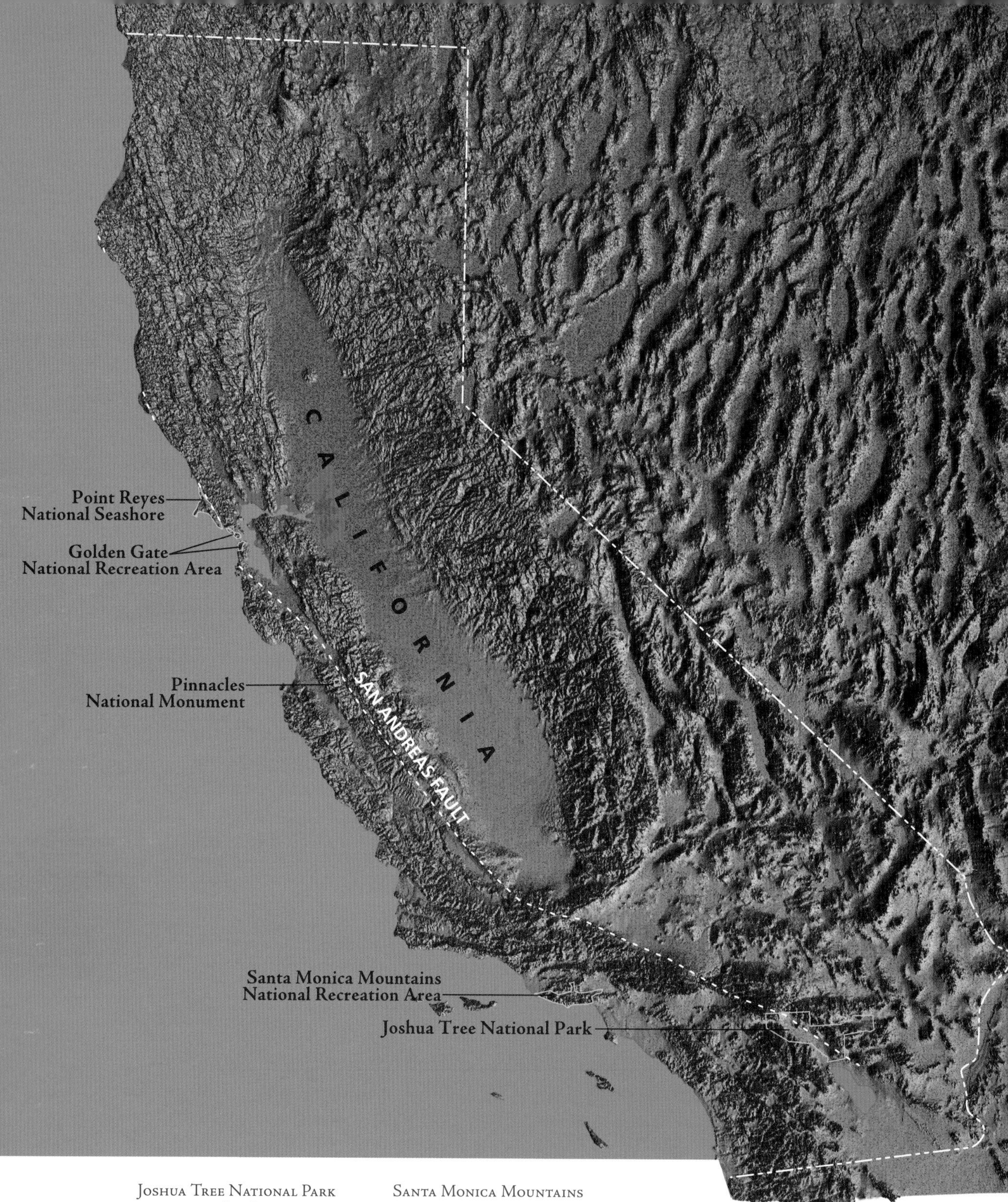

Joshua Tree National Park
74485 National Park Drive
Twenty-nine Palms, CA 92277
760/367-5500

Pinnacles National Monument
5000 Highway 146
Pacines, CA 95043
408/389-4485

Santa Monica Mountains National Recreation Area
401 West Hillcrest Drive
Thousand Oaks, CA 91360
805/370-2301

Golden Gate National Recreation Area
Building 201, Fort Mason
San Francisco, CA 94123
415/556-0560

Point Reyes National Seashore
Point Reyes Station, CA 94956
415/663-1092

ParkNet *(National Parks on the Internet):*
http://www.nps.gov

beyond. That term may be used in a comprehensive sense, significant of the genetic and structural unity which runs thru them. The map accompanying his text shows a single fault—the San Andreas—stretching from Punta Gorda on the north through San Gorgonio Pass. Current maps would extend his definition only seventy miles farther southeast to brush the Salton Sea.

The preponderance of California's population lives within the grasp of the San Andreas and its subsidiary faults. San Francisco and the Bay Area were rocked again in 1989 during the World Series. And, like it or not, life in Los Angeles and its suburbs is one long earthquake drill. And yet, how many Californians can point exactly to the fault segment closest to their home or work? Do many of the daily commuters on the Junipero Serra freeway between San Jose and San Francisco know of the fault's immediate proximity? How many motorists whizzing home through Cajon Pass on Interstate 15 realize that they are crossing from the North American to the Pacific Plate?

California is fortunate to have five national parks, monuments, or recreation areas that incorporate land shaped by the San Andreas. Each offers access to a unique view of the fault as it cuts through the state. Joshua Tree National Park sits above the southeastern end of the San Andreas Fault as it emerges from the Transverse Mountains east of Los Angeles. Driving into the park from the north, you cross the Pinto Mountain Fault, an important subsidiary of the San Andreas. This fault is dotted with palm oases, places where groundwater pushes up through crumbled rock toward the surface. The park's Geology Tour Road leads down to an impressive view of the Blue Cut Fault. Motor south through the light-colored White Tank monzogranite that makes up the bulk of the park, and on through the darker Pinto gneiss until you reach Keys View. Be prepared to sit a spell: the vista is as captivating as it is complex. San Gorgonio and San Jacinto peaks loom above Palm Springs and the Coachella Valley. The Salton Sea lies thirty miles to the south, fifty-three hundred feet below. And the San Andreas rips diagonally across the entire scene. Joshua Tree sits on the western edge of the North American Plate. Every building, roadbed, valley, and mountain that you see on the other side of the San Andreas is riding on the Pacific Plate, moving inexorably to the northwest.

Pinnacles National Monument, south of San Benito, lies midway up the San Andreas Fault in central California. From the top of its High Peaks Trail, the San Andreas is visible 10 miles to the east. The crags and spires here are volcanic rocks, spewed out twenty-three

Volcanic rocks along the High Peaks Trail, Pinnacles National Monument

million years ago as an incandescent series of andesite, dacite, rhyolite, and pyroclastic flows. Volcanic rocks like these are relatively scarce in coastal California. If you go 195 miles to the southeast, to Antelope Valley a few miles east of Tejon Pass, you will find the Neenach Volcanic Formation. These rocks offer a precise stratigraphic match with Pinnacles that is impossible to ignore: same rocks, same age. The two locales must once have been adjacent. The Neenach Volcanics sit solidly aboard the North American Plate, while Pinnacles National Monument, west of the San Andreas, rests firmly on the Pacific. Geologists have yet to uncover more compelling evidence of the magnitude of movement along the San Andreas.

Santa Monica Mountains and Golden Gate national recreation areas (in or near Los Angeles and San Francisco, respectively) are both urban parks on or adjacent to the San Andreas. In southern California, Kanan Dume and Malibu Canyon roads traverse the wildly tilted sandstones and volcanic formations that have been thrust skyward within the Santa Monica Mountains National

Malibu State Park within the Santa Monica Mountains National Recreation Area, Los Angeles

Recreation Area. These mountains are a spectacular example of the Transverse Ranges, created as the Pacific Plate slides past and, in places like this, into the North American Plate. Uplift of the Santa Monica Mountains continues today at a gallop: one inch every thousand years.

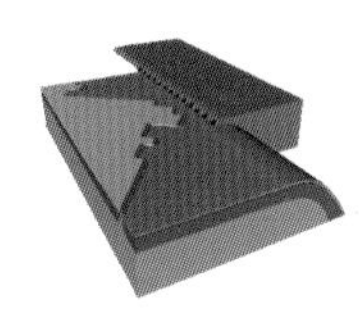

Golden Gate National Recreation Area is a loose collection of intimate parklands bracketing San Francisco, all of which are within a stone's throw of the San Andreas Fault. From the Ocean Beach overlook at Fort Funston in San Francisco, you can see Mussel Rock, where the San Andreas slips offshore on its way north to Bolinas Bay. Across the Golden Gate Bridge, the Marin Headlands offer a dramatic view of San Francisco and an intimate look at the Franciscan Formation. These twisted cherts hint of the tortured creation of this part of the continent.

Point Reyes National Seashore is twenty miles north of San Francisco, past Stinson Beach and Bolinas Bay. Southerly access to the park is along California Highway 1, a road that clings to the western brink of North America with varying degrees of success.

Ocean Beach and Mussel Rock from Fort Funston, Golden Gate National Recreation Area

The open lands and pristine seashores of Point Reyes are aboard the Pacific Plate. A well-signed Earthquake Trail strikes out from the Bear Valley Visitor Center near Olema, exploring the soft hummocks and now-smooth scarps along which the ground was torn in 1906. Much of Point Reyes is underlain by layers of rocks called the Monterey Formation, deposited fifteen to sixty million years ago when this part of the Pacific Plate was still sitting one hundred miles to the southeast, about where Monterey is now.

At each of these parks, we see rocks tilted crazily or uplifted bodily; we see rocks from the same source separated by almost two hundred miles; we see good rocks—government property—torn, folded, and mutilated. We need a framework within which to understand these diverse changes. It has taken geologists a century to realize how the San Andreas Fault links all of these beautiful and twisted landscapes.

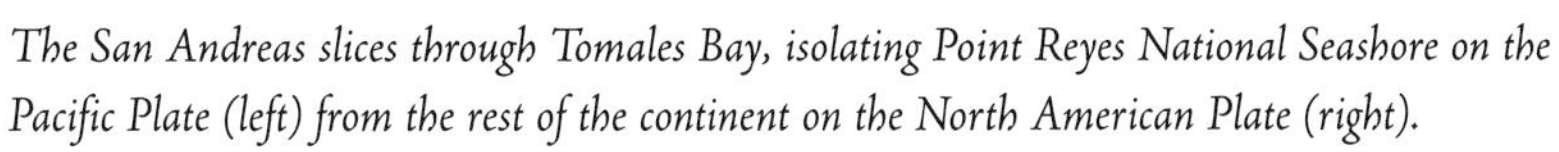

The San Andreas slices through Tomales Bay, isolating Point Reyes National Seashore on the Pacific Plate (left) from the rest of the continent on the North American Plate (right).

Plate Tectonics

THE GREATEST ADVANCES MADE in any science are neither linear nor predictable. People chug along for centuries, tinkering here and there with a few new or interesting ideas, but basic interpretations of the natural world remain pretty much unchanged. Then someone comes up with a new paradigm and suddenly, everything is turned upside down.

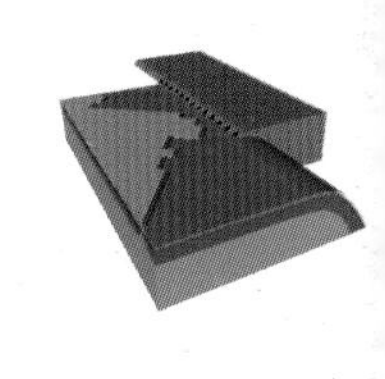

Charles Darwin revolutionized biology with the theory of natural selection. Niels Bohr forever changed physics when he introduced his ideas about the atom. Cell physiology leapt into the twentieth century when Watson and Crick revealed the intriguing beauty and elegance of the DNA molecule. These ideas all had deep roots in the thinking of earlier biologists, physicists, and physiologists. But Darwin, Bohr, and Watson and Crick each dismantled an established science and then put it back together in a fundamentally new way. Their theories resulted in a sweeping reorganization of other scientists' understanding of how the world fits together.

When the basic tenets of geology were formalized in the eighteenth and early nineteenth centuries, the earth was considered a solid sphere of rock. Seas might come and go as the land rose and

---

Sag ponds along the San Andreas, south of California Highway 46

fell, but the continents were thought to be fixed in place, never moving relative to one another. To be sure, there was much to be learned even within this static view of the world. Early geologists described rock types and puzzled over the surface processes of erosion and deposition; they studied the variety of life forms in the fossil record. They slowly began to grasp the awesome extent of geologic time, relying upon gradual but steady incremental change that eventually created the complex landscapes within which we live—"eventually" on the order of millions or even billions of years. Various theories were advanced to explain the sinking of sea floors and rising of mountain chains, but to be honest, none were very satisfying.

Geology experienced its earth-shaking revolution in the 1960s with the introduction of the theory of plate tectonics. Seeds of this theory had been planted as early as the sixteenth century, when mariners and map-makers remarked upon the uncanny fit of Africa against South America—if one could just remove the Atlantic Ocean. The significance of this observation languished until 1912, when a young German meteorologist published two articles that introduced his concept of continental drift. Alfred Wegener proposed that until two hundred million years ago, all continental landmasses were

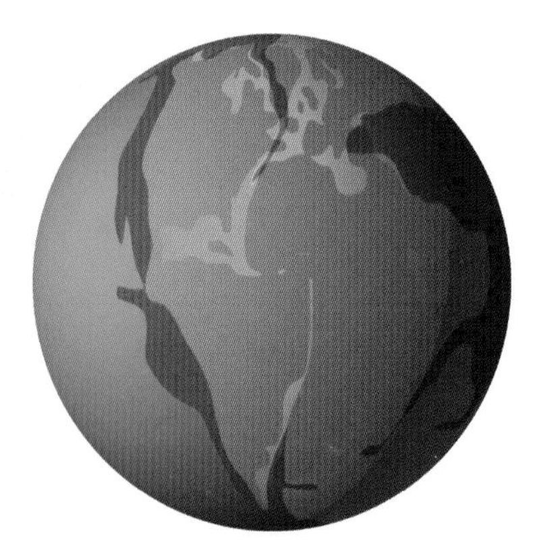

Permian
225 million years ago

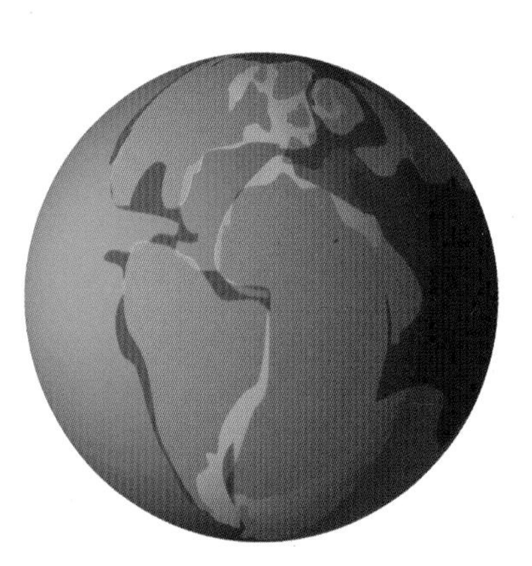

Triassic
200 million years ago

Jurassic
135 million years ago

Pangaea was a single large landmass that existed in the Permian period. It subsequently began to fragment and the pieces gradually moved into their present positions. (After Kious and Tilling, This Dynamic Earth.*)*

clustered into a single super-continent, which he dubbed Pangaea. His super-continent was then torn apart and the land masses that we now call continents sailed off to their current positions.

Wegener based his theory on observations of identical fossil assemblages on continents that are now separated by thousands of miles of ocean. He pointed to environments of sediment deposition—desert sand dunes, glacier-derived conglomerates, tropical swamps—recorded in rocks that today lie at latitudes that couldn't possibly be so dry, cold, or hot. But Wegener died long before his ideas were accepted. The geologic community refused to believe that continents had plowed around the globe like icebreakers through solid ocean floors. After all, what force could possibly accomplish this?

As early as 1855, bathymetric ocean charts had shown an underwater mountain chain in the middle of the Atlantic. Sonar, used extensively in World Wars I and II by surface ships searching for submarines, expanded this view. Harry Hess (a professor of geology at Princeton when WWII broke out) captained an assault transport to battles in the Marianas and Iwo Jima. Between skirmishes, he used his depth-sounder to map significant portions of the Pacific

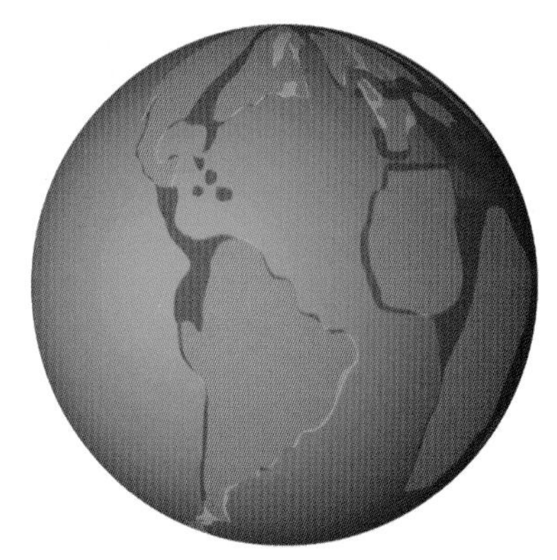

Cretaceous
65 million years ago

Present

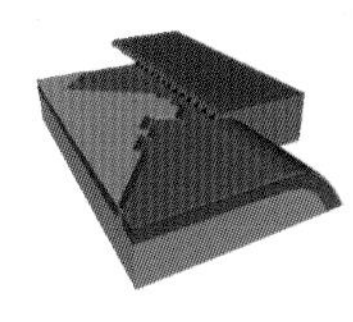

Ocean floor. Hess discovered an underwater world of jagged peaks, deep trenches, and ocean-spanning mountain chains. He also knew of Wegener's continental-drift hypothesis. After the war, armed with his seafloor maps, Hess expanded Wegener's ideas into the theory of *seafloor spreading,* which was first published in 1962.

Hess postulated that seafloor rocks are created when molten lava *(magma)* wells up from the earth's interior at a mid-oceanic ridge. This magma cools and solidifies to form new ocean floor.

Mid-oceanic ridges run as nearly continuous seams that stitch together all of the world's ocean floors. These ridges are the hot, young spreading centers where fresh magma floods to the surface to become newly minted basalt.

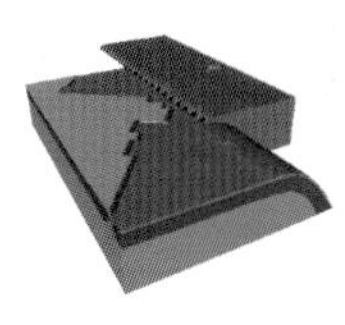

The sea floor then spreads passively away from the ridge until it is swallowed into trenches hundreds or thousands of miles away—essentially, a series of global conveyor belts. This theory went a long way toward explaining a curious observation: sediments dredged from ocean bottoms around the world were surprisingly young—none were older than 180 million years, despite the fact that continental rocks have yielded fossils three and four times as old. Metamorphic rocks found in Greenland have been dated at 3.8 billion years (close to our best estimates of the earth's age of 4.6

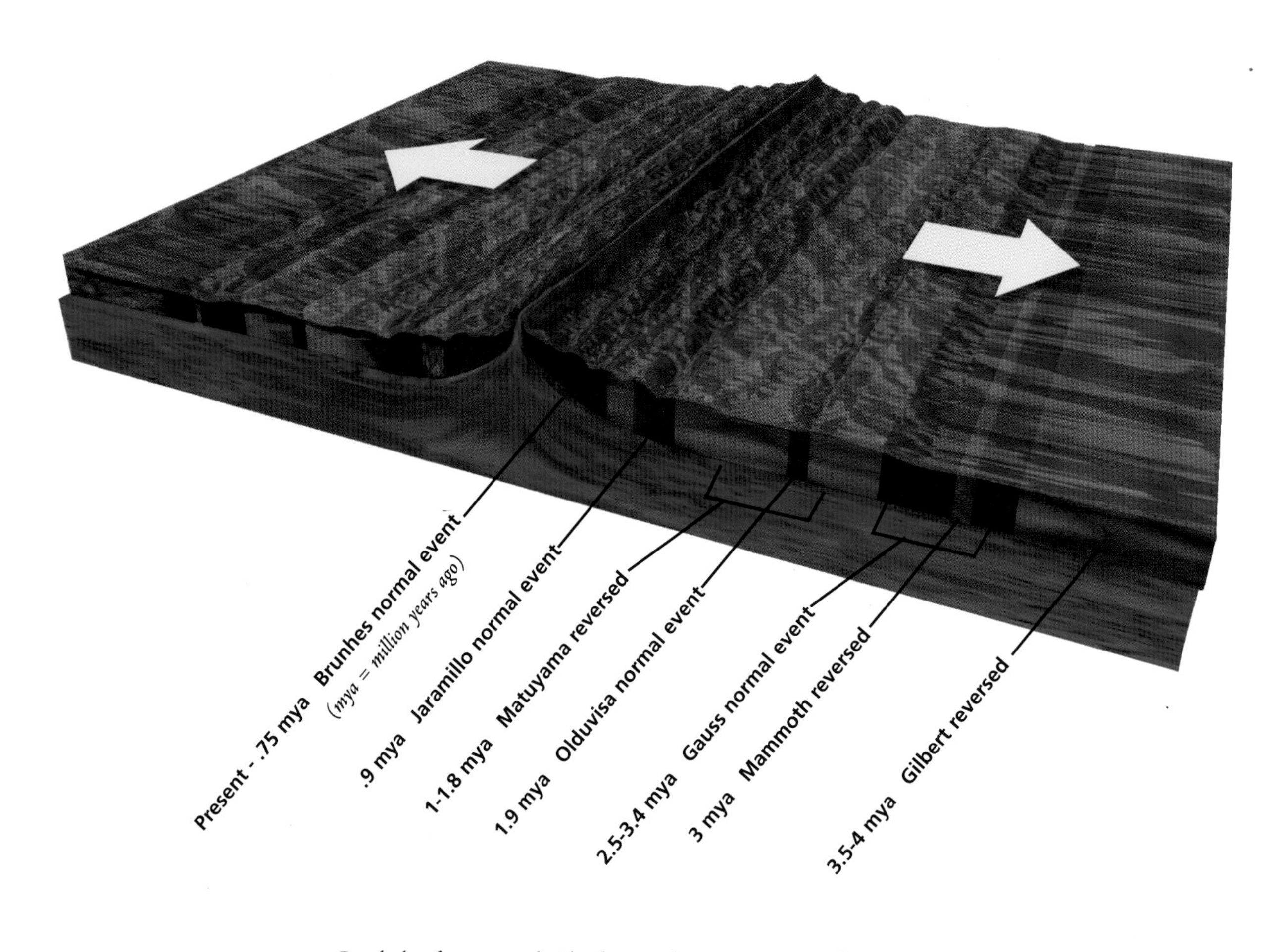

Basalt that forms on each side of a spreading center assumes the magnetic orientation that exists when it first cools. Since the earth's magnetic poles flip back and forth every few hundred thousand years, basalts will be imprinted either with a normal (north pointing north) or reversed (north pointing south) orientation.

billion years). Hess argued that ocean floors must be consistently younger than the continents because, unlike continents, the ocean floors are being continuously recycled. Geologists were intrigued but still not convinced.

Another essential clue was uncovered in the mid-1960s by scientists studying magnetization of rocks. The earth's magnetic field is strong enough to imprint itself upon susceptible iron-bearing minerals such as magnetite. Thus one can measure magnetism within a grain of rock and know the orientation of the earth's field when that rock was formed.

Navigators and scientists have long known that the earth's magnetic field wobbles a bit. Wobbles notwithstanding, throughout recorded history, the north arrow on a compass has always pointed at least roughly toward the Arctic. At the beginning of the twentieth century, scientists working first in France and then in Japan discovered that some rocks record times of reversed magnetic polarity. A compass's north arrow would have pointed south during these periods! The implications were staggering: something inside the earth's core must periodically change. The earth could no longer be considered a static chunk of rock. Not only must there be an engine able to create this magnetic field deep beneath the surface, but that engine must be capable of dynamic variations that would account for the field's reversals. These polarity reversals have happened randomly throughout geologic history, occurring about once every two hundred thousand years when averaged over at least the last forty-five million years.

Who cares? Vine, Matthews, and Morley did. In 1962, the US Naval Oceanographic Office published its catalog of ocean-floor magnetic polarities. English geologists Frederick Vine and Drummond Matthews and Canadian Lawrence Morley sifted through this data and arrived at a remarkable conclusion: magnetic polarity is arranged along the ocean floor in stripes distributed symmetrically around the mid-ocean ridges. Hess had hypothesized that new ocean-floor rocks are created at these ridges. Vine, Matthews, and Morley went on to demonstrate that rocks on either side of a ridge were formed simultaneously, both sides recording identical histories of magnetic polarity; each side was then transported equal distances from the ridge.

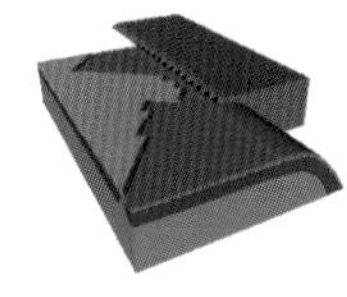

The theories of continental drift and seafloor spreading were converging toward what would become our modern theory of plate tectonics. In its simplest form, this theory tries to explain movements of the earth's crust driven by motion within a semi-solid mantle. According to plate tectonics, the earth is made up of a dense inner core; a somewhat lighter *mantle;* and a thin, stiff *crust.* The core is composed of a solid inner sphere and a liquid outer shell, with a combined thickness of 2,166 miles. Rocks within the mantle, which is about 1,800 miles thick, can be bent and twisted as though they were taffy. The thin and comparatively rigid crust varies in thickness from as little as 4 miles when measured beneath the ocean floor to as much as 30 miles when measured beneath typical dry continental land.

The crust, then, is the thin skin of this peach we call earth. Immediately below—at depths of fifty to about seventy-five miles—is a region of the mantle that behaves as a viscoelastic material, still stiff enough to be more elastic than viscous. This region, along with the overlying crust, is called the *lithosphere.* The theory of plate tectonics postulates that the lithosphere is made up of at least a dozen discrete plates, some continental, some oceanic, and some containing both continental and oceanic crusts sutured together. Each plate moves over the face of the earth as a more-or-less intact unit. When a collision with another plate occurs, the plate's interior is less likely to be disturbed than its edge. While Wegener was ridiculed because he couldn't suggest a force capable of driving continents around the

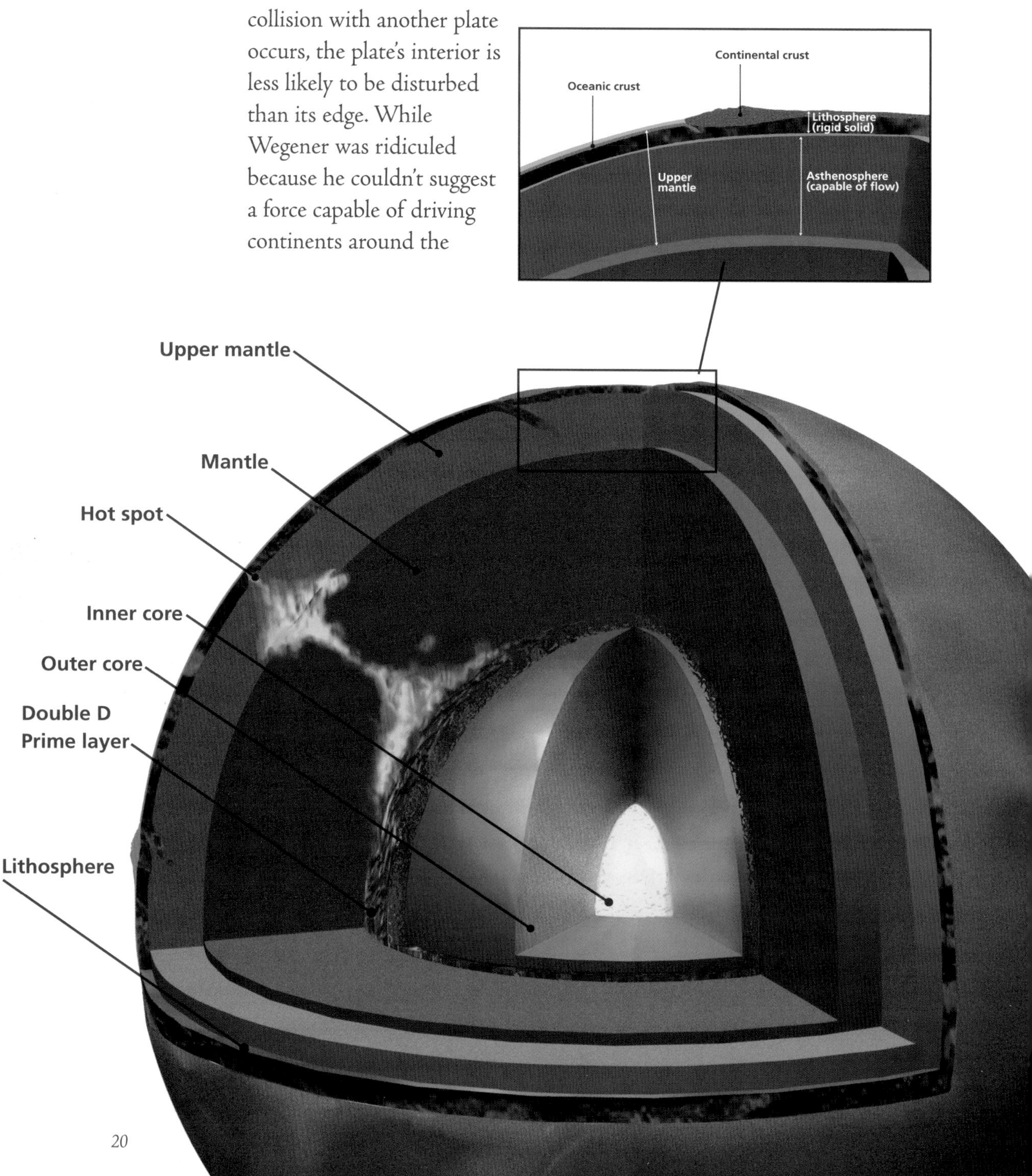

globe, geologists now believe that the hot mantle contains convective cells of circulating magma, not unlike a pot of boiling soup. Magma wells up toward spreading centers at ocean ridges and cools to become ocean-floor rocks, which are carried laterally and then dive back down into the mantle. Plates move because they are the top of a circulating mantle cell.

Oceanic crust is made of heavy basalt and gabbro that form at the mid-ocean ridges. Continental crust is lighter, composed of rocks such as quartz-rich granite and sandstone. Basalt, volume for volume, is 2.9 times heavier than water, while granite is 2.7 times and sandstone, 2.4 times as heavy. These density differences may *seem* trivial. But continents, with their granite and sandstone, float higher on the underlying mantle than ocean floors. And that makes all the difference.

Plates cruising around the world can interact in one of only three ways. They can diverge, collide, or slide past one another. Plates diverge at underwater spreading centers such as the Mid-Atlantic Ridge that bisects the North and South Atlantic oceans. When magma first surfaces, it is hot, swollen, and buoyant; with time and distance from the ridge, it cools, shrinks, and becomes more dense. The warm, lighter ridges stand much higher than the cooler surrounding ocean floor. Continents can also spread apart, with basalt flooding in to fill the gap. The United States is torn along a number of such breaks: the Rio Grande Rift running north-south

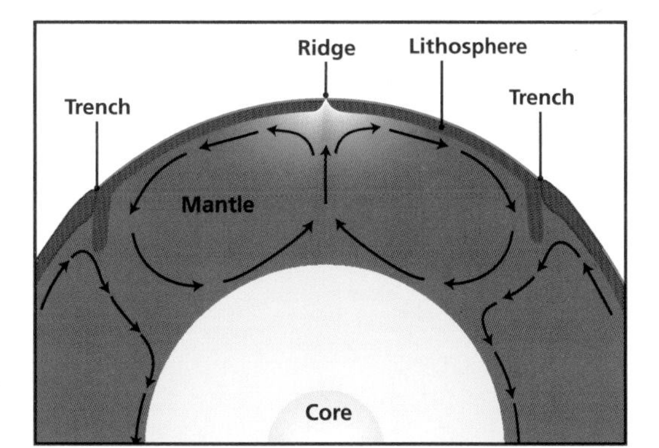

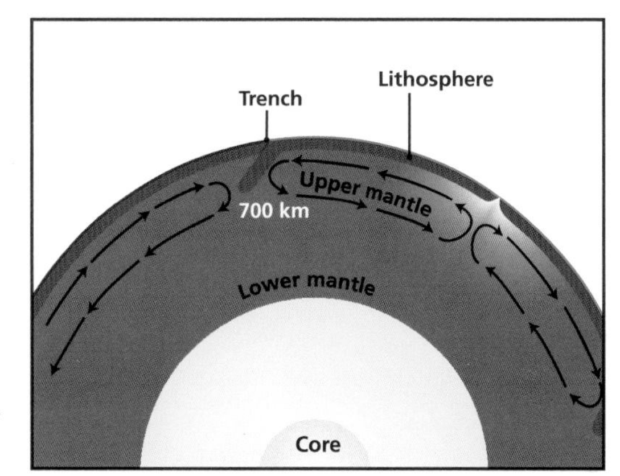

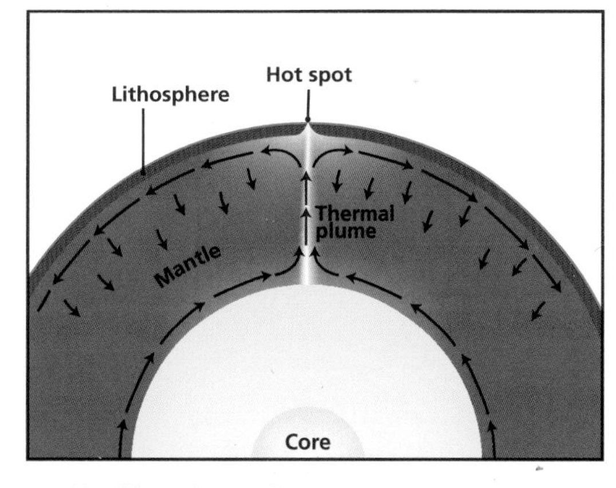

Models of forces driving plate tectonics
Top: *Large convection cells in the mantle carry the lithosphere in conveyor-belt fashion.*
Middle: *Convection cells, confined to the mobile upper mantle.*
Bottom: *A hot plume suggests that all upward movement is confined to a few narrow plumes, while downward flow occurs slowly throughout the remaining mantle segments.*

PLATE CONVERGENCE

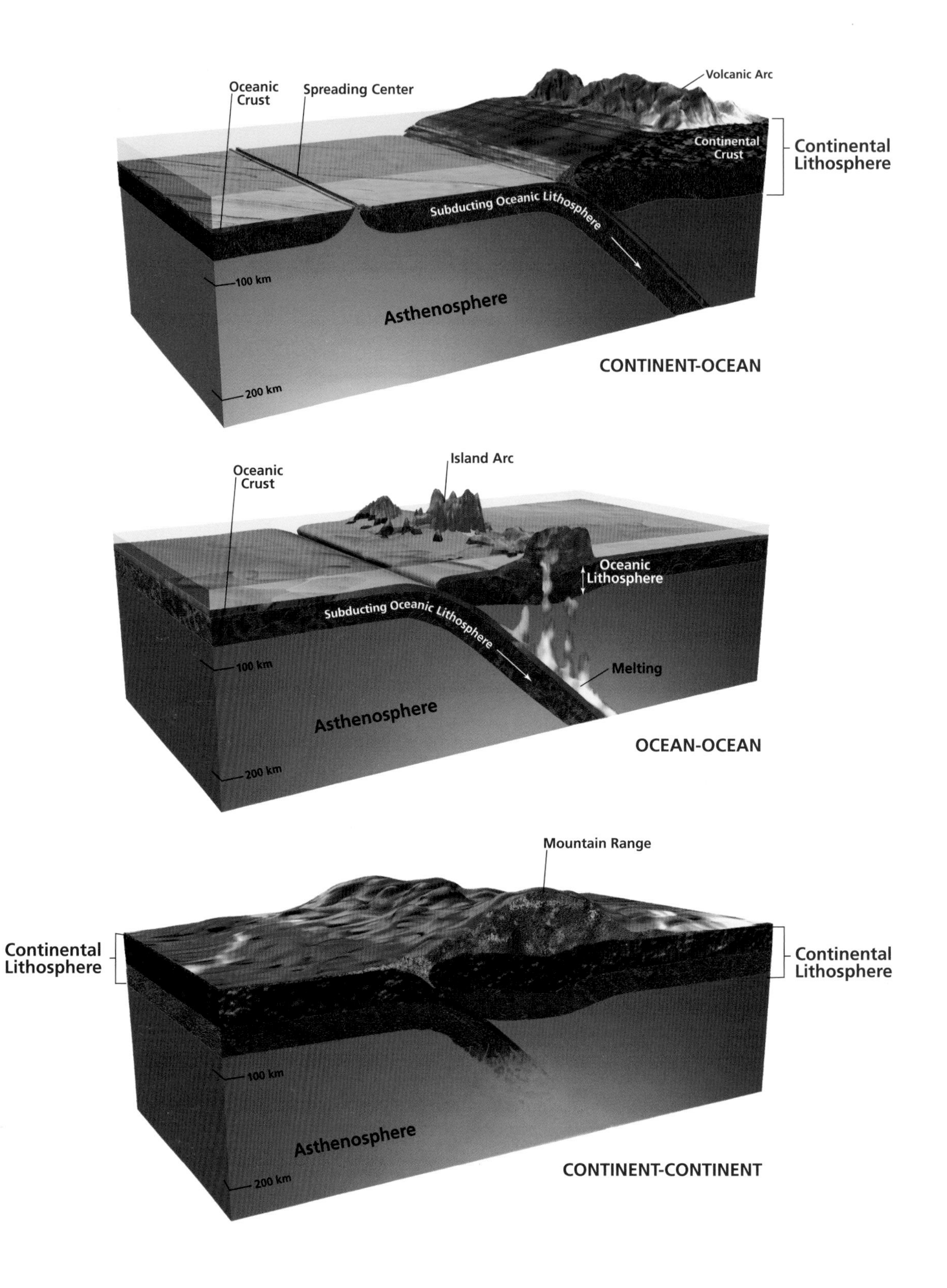

through New Mexico, and the hot crust beneath California's Salton Sea, stretched thin as its two sides drift apart.

Collisions also come in three flavors: continent-ocean, ocean-ocean, or continent-continent. In the first case, the heavier and thinner oceanic crust is doomed to be overridden by the lighter continental crust. Continental rocks are preserved at the surface, but the ocean floor *subducts,* or dives back into the mantle, in a subduction zone. This zone typically angles downward at about forty-five degrees. The cold subducting crust initially acts as a block, cracking like solid rock rather than flowing like liquid. We feel this cracking as earthquakes. As the cool slab of ocean floor descends, it is heated by the surrounding mantle and begins to melt. Geophysicists have mapped zones of earthquakes down to depths as great as 435 miles. Presumably, subducted slabs are so softened beneath this level that they can no longer behave as brittle solid rock and generate earthquakes. Instead, they bend and flow rather than break. Some lighter quartz-rich constituents are melted and scraped off the subducting plate, floating back to the surface to become volcanoes—Mount Saint Helens—or crystallize underground to become granite—the Sierra Nevada.

When two oceanic plates collide, one subducts beneath the other. This spawns volcanic island arcs such as Japan, or Alaska's Aleutian Islands. When two continents collide, neither is dense enough to subduct beneath the other. Instead, their sutured crust is pushed upward into towering mountains like the Himalayas, created by India's collision with Asia that began fifty million years ago and continues today. The thickened crust bows down into the mantle, resulting in continental plate thicknesses that can approach fifty miles.

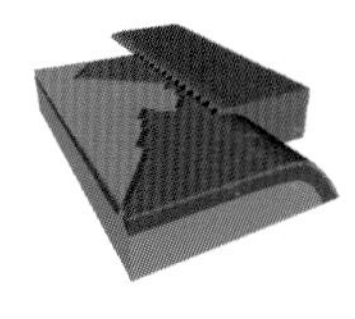

So we have plates twirling across the surface of the earth like interlocked lily pads on a spherical pond. Sometimes they collide and sometimes they pull apart, leaving a heaped-up trail of mountains here and a wrecking yard of subduction zones there. With a great deal of sleuthing, we can read their prints in the sand and track relative plate motion back through geologic time.

But is it possible to determine the absolute motion of a single plate in relation to the underlying mantle? To do so, we need markers not only on the earth's surface, but also within its interior. A string of islands stretching across the Pacific offers insight into this question. The Hawaiian Islands are all volcanic;

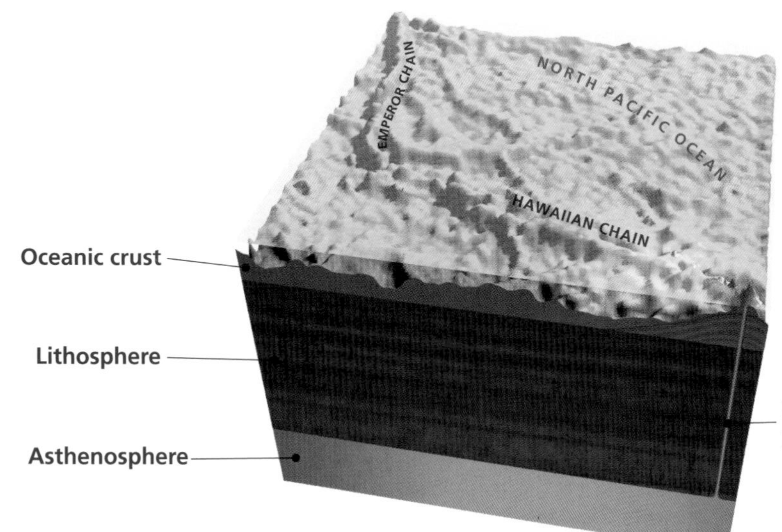

The Hawaiian Islands are the most recent evidence of volcanic eruptions that stretch in a long chain northwest across the Pacific Ocean. The tops of most of the volcanoes are now underwater, but a few—such as Midway Island—just reach the surface. This chain marks the passage of the Pacific Plate over a stationary hot spot within the mantle. The chain's pronounced angulation was produced 43 million years ago when the plate changed directions.

indeed, magma intermittently erupts on the Big Island every few years, running in molten rivers to the sea. The Big Island of Hawaii is the youngest; as the chain is traced to the northwest, the islands become progressively older. Kauai, the most northwesterly, is 5.5 million years old. Midway Island, fifteen hundred miles farther still to the northwest, is built of 27.2 million-year-old basalt. Sonar maps of the Pacific floor reveal a continuous line of underwater seamounts between Hawaii and Midway, all volcanic, each older than the one that precedes it.

Imagine a welding torch fixed beneath a steel plate that is being drawn across the torch's flame. The stationary torch leaves a seared trail of welded steel on the bottom of that moving plate. The Hawaiian Chain, stretching from the Big Island well past Midway Island, is a lot like that welded line. And the Pacific Plate is a lot like the sheet of steel. For at least the last seventy million years, there has been a point source of magma fixed within the mantle beneath the Pacific Plate. As the plate drags over this hot spot, magma cuts through the Pacific Plate and erupts to form one basaltic island after the next. Knowing the distance between the islands and the ages of the basalt, we can deduce that the absolute motion of the Pacific Plate has been 3.5 inches per year to the northwest.

So the plates are moving to (convergence) and fro (extension), relative not only to one another, but relative also to the underlying mantle. One more type of motion—the one pertinent to the San Andreas Fault—remains to be explained. Plates, which can collide or diverge, also have the option of just sliding past one another. The Mid-Atlantic Ridge and the East Pacific Rise are parts of what at first appears to be a continuous forty thousand-mile-long rift encircling the globe. But examined in greater detail, this rift turns out to be a series of discontinuous spreading centers connected by faults that accommodate subsidiary lateral movement. J. Tuzo Wilson first proposed that rocks could skid horizontally along these breaks between spreading centers. He called these connecting breaks *transform faults*, because they "transform" divergence from one spreading center to another. These faults not only connect the spreading center segments, but also stretch beyond as fracture zones. Fracture zones accommodate the differential cooling and contraction of ocean-floor basalt of varying ages.

Transform faults are associated with spreading centers, and thus most are under oceans. A few, however, have found their way onto dry land. The San Andreas Fault is a perfect example. Here, along this fault, the eastern edge of the Pacific Plate has shaved off a sliver of continental crust that now glides along the western edge of North America.

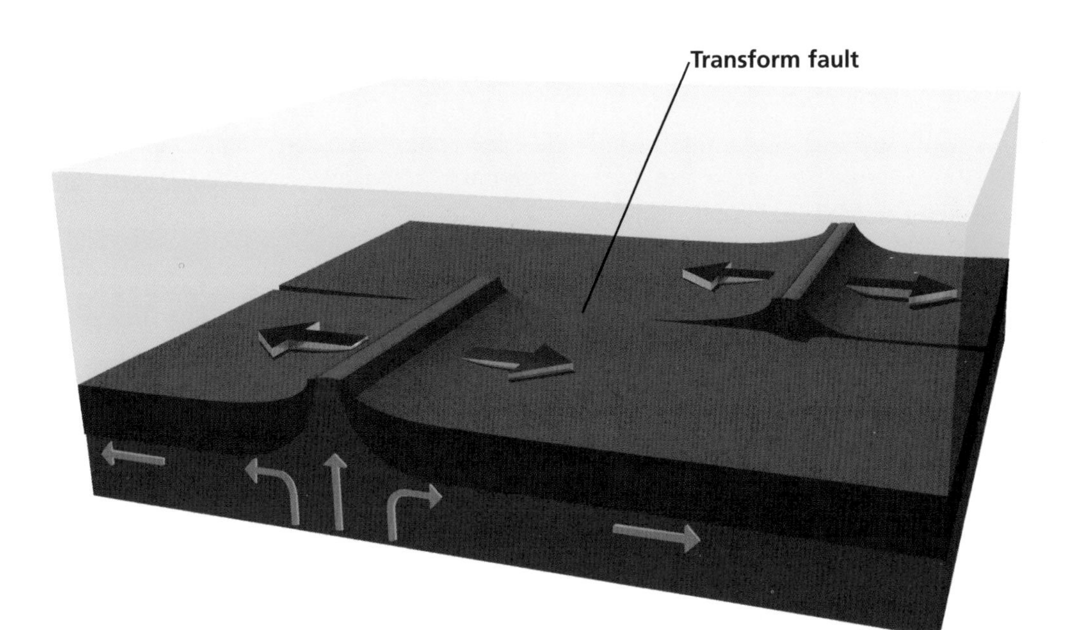

Spreading centers are straight lines imposed on a curved earth; in places, the centers must be offset to accommodate the earth's curvature. They are then connected by transform faults, where crust formed by one spreading center slides laterally past crust formed by a neighboring spreading center.

A Land in Motion

THE SAN ANDREAS FAULT has not always sliced through California as it does today. Indeed, sometime before about sixteen million years ago, there was no San Andreas Fault at all. So how did it come to be? There are just three players in this particular geologic drama: the Pacific and North American plates, as might be expected, and something called the Farallon Plate, now almost totally hidden from view. Let's back up and get a running start; back, say, a hundred and fifty million years ago.

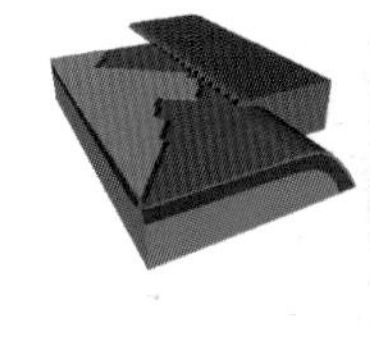

Alfred Wegener's Pangaea had broken up. The Pacific—not just the plate, but also the ocean itself—was much larger than it is today. The East Pacific Rise was a seafloor spreading center out in the middle of the Pacific Ocean, stretching thousands of miles north and south. This rift separated the Pacific Plate (which extended to the northwest) from the Farallon Plate (which extended to the southeast). Both were oceanic plates: thin and relatively dense. The

Offset streams, Carrizo Plain

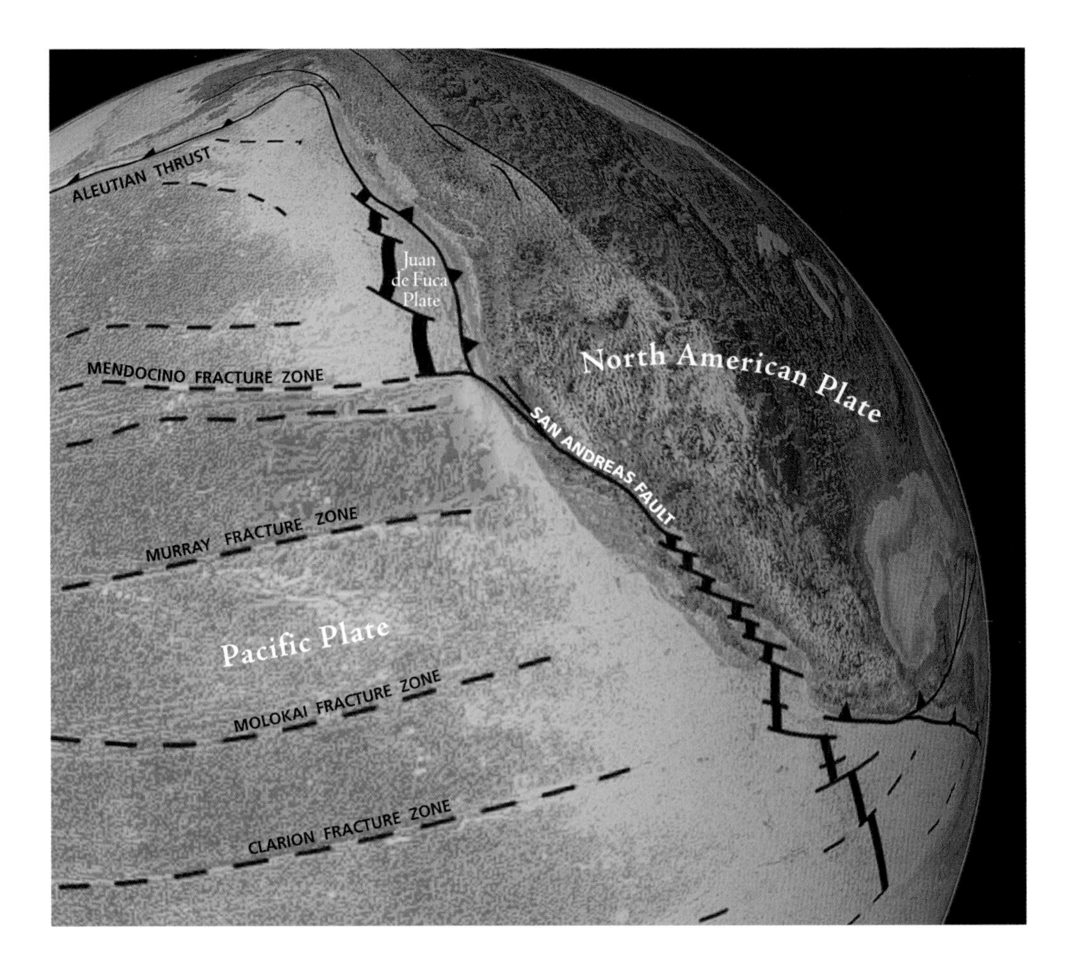

North American Plate, springing from its starting block in Pangaea just fifty million years earlier, was racing westward, overrunning the far eastern end of the Farallon Plate. As the Farallon Plate was subducted, all sorts of fireworks were going off: creation of the Sierra Nevada granite; growth of the Rocky Mountains; and eventually, the rise of the Colorado Plateau. Without the subduction of the Farallon Plate beneath the North American Plate, the American West would be every bit as geographically exciting as Kansas.

Life became a little more complicated about twenty-eight million years ago, when the North American Plate first overran portions of the East Pacific Rise. As the spreading center disappeared down the subduction zone, the North American Plate (moving west) encountered the Pacific Plate (moving northwest). As a result, interplate motion was transformed from east-directed subduction of the Farallon Plate under the North America Plate to northwest-directed sideways (or transform) movement of the Pacific Plate relative to the North American. Sideways, or *strike-slip*, motion first occurred along

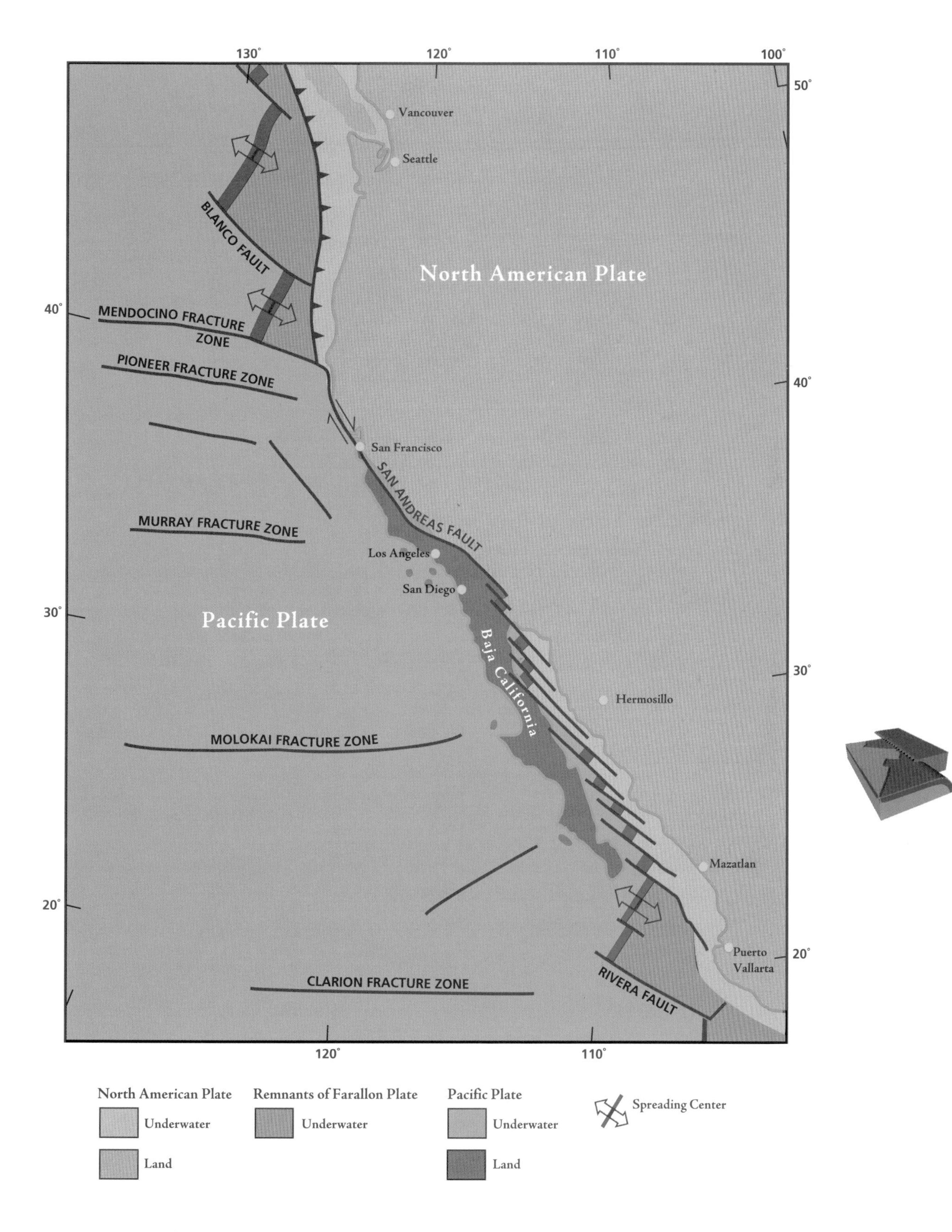

San Andreas Fault as the boundary between the North American, Farallon, and Pacific plates

NORTH AMERICAN PLATE
PACIFIC PLATE
FARALLON PLATE
40 million years ago

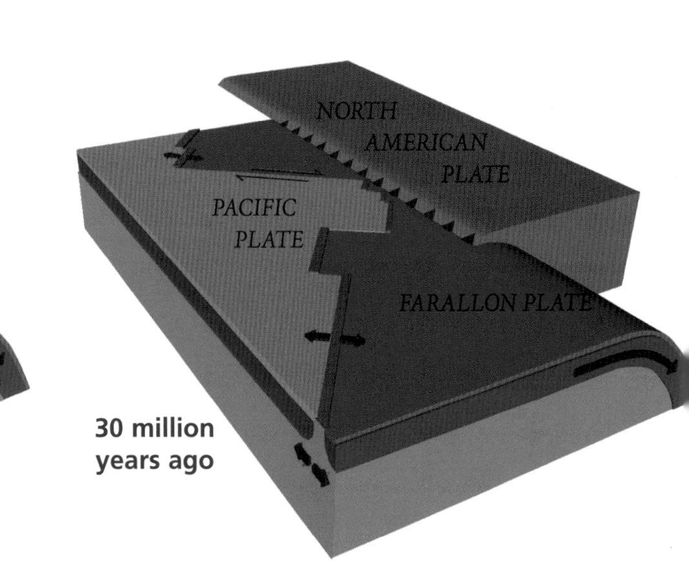

a number of precursors to the San Andreas, such as the San Gregorio-Hosgri Fault near Santa Cruz and the San Gabriel Fault just north of Los Angeles. The San Andreas Fault itself first began to move sometime before sixteen million years ago. Over the next few million years, as other older faults slowed down, the San Andreas gradually became the dominant plane of strike-slip motion between the Pacific and North American plates.

Today this strike-slip zone stretches from a point just off the Mendocino coastline south through most of California to the Rivera Triple Junction near Mazatlan in Mexico. The San Andreas Fault makes up the northern half of this zone from the Mendocino Triple Junction to the Salton Sea. A series of other transform faults connects small spreading centers from the Salton Sea through the Gulf of California to the Rivera Triple Junction, and constitutes the southern half.

The "triple junctions" are so called because they are points where three plates meet. North of the Mendocino Triple Junction, the strike-slip motion between the North American and the Pacific plates is replaced by subduction of a remnant of the Farallon (called the Juan de Fuca Plate) beneath North America. Similarly, south of the Rivera Triple Junction, Pacific-North American strike-slip contact is replaced by subduction of another fragment of the old Farallon Plate—here known as the Rivera Plate.

I WOULD BE THE FIRST TO ADMIT that this spinning and whizzing of plates is confusing. An analogy may help. I submit that, as modern travelers, our greatest fear doesn't necessarily involve becoming that silently screaming face in the window of an airplane

Plate Subduction

To see the plate movement as shown in these five diagrams, watch the small illustration along the page edge as you fan the pages.

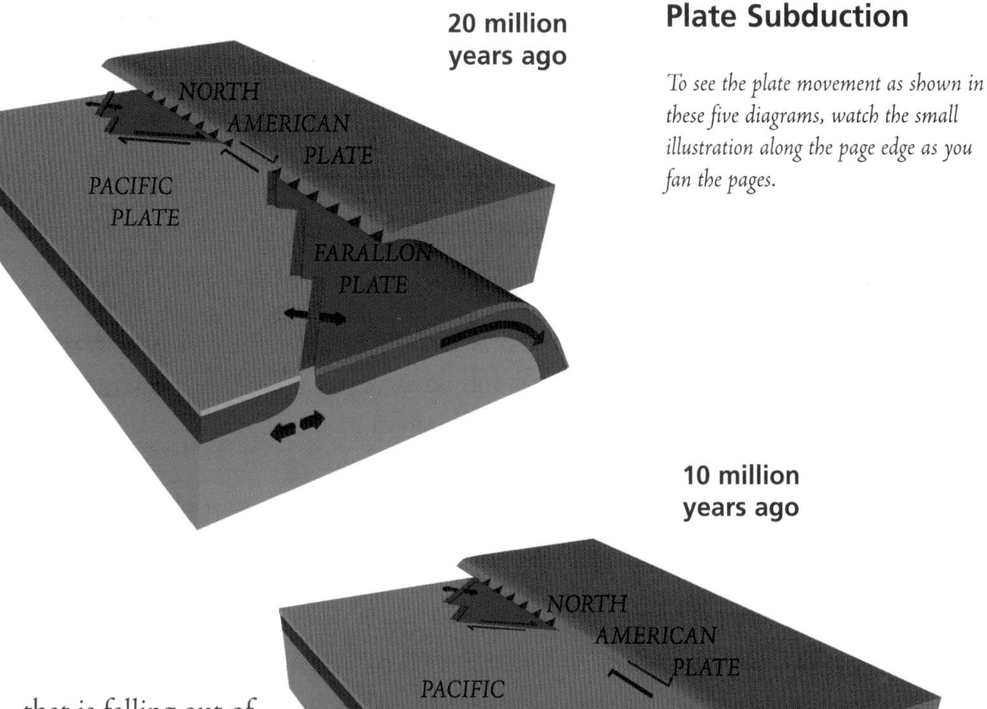

10 million years ago

NORTH AMERICAN PLATE

PACIFIC PLATE

that is falling out of the sky. Nor is it even—heaven forbid—the horror of losing one's luggage between connecting flights. Rather, our greatest fear is of getting a foot stuck in the airport's moving sidewalk. I am convinced that the entire animal kingdom acquired this atavistic anxiety over hundreds of millions of years as we pranced around trying to avoid subduction zones.

Let's tempt fate. Let's stand on that moving sidewalk. I can walk as the sidewalk moves, decreasing the time left until I get to the end of the sidewalk if I walk straight ahead, or increasing the time left if I walk away from the end. And if I walk away at exactly the right speed, it might appear to an observer that I am standing in place.

PRESENT

NORTH AMERICAN PLATE

PACIFIC PLATE

Here's the trick: assume that the moving sidewalk is quite broad, not just three or four feet, but a couple of thousand miles wide. And now, rather than walking straight away from the end, I will walk away at a forty-five-degree angle. At just the right speed, I could not only avoid encountering the end of the sidewalk, but could also move right or left relative to it. To a person standing just off the end of the sidewalk, it would appear that my motion is sideways rather than forward or backward. If I slow down a little bit, I will still move laterally, but will gradually be dragged toward the inevitable. Remind me to tie my shoelaces.

Coastal California is on its own moving sidewalk. The Pacific Plate (including the continental sliver of California that now lies west of the San Andreas) is steaming northwestward just fast enough to keep from being subducted. The result is that, rather than being stuffed beneath Joshua Tree National Park, Palm Springs is destined to eventually slide past San Francisco.

LET'S CONTINUE OUR LOOK BACK IN TIME. The West Coast was first shaped by that subduction zone along which the oceanic Farallon Plate dove beneath the westward-advancing North American Plate. Embedded within the Farallon Plate were odd slivers of buoyant rock—a seamount here, an odd continental scrap there—riding high on this oceanic crust. The slivers' origins are anyone's guess; perhaps they are the Farallon's battle scars from prior encounters with other, even older, plates. As the plate subducted, these slivers were scraped off and plastered against the leading edge of North America. We now know these slivers as accreted terranes—fault-bounded territories, each with a unique geologic heritage clearly different from its neighbors. Southern California is an amalgam of some ten different terranes, many of which had congealed into one ocean-going microplate before being accreted to North America approximately fifty-five million years ago. The Franciscan Complex, now the bedrock of much of coastal northern California, is a cartel of still more terranes, some of which appear to have migrated up from the Southern Hemisphere. All of these terranes rafted with the Farallon Plate (and possibly other earlier plates, such as the fearsome Plate X) into North America, and were preserved by accretion rather than being sucked down into the subduction trench.

Meanwhile, prior to thirty million years ago, the East Pacific Rise obliquely approached the California coast. Twenty-eight million years ago, the it finally bumped into and then slid under the North American Plate near the current latitude of San Diego or northern Baja California. For the first time, the Pacific Plate was in direct contact with the North American Plate. As the East Pacific Rise dove beneath the North American, its production of new oceanic crust presumably fizzled out.

California is a mosaic of many different terranes, each accreted to the continent from a different source and at a different time.

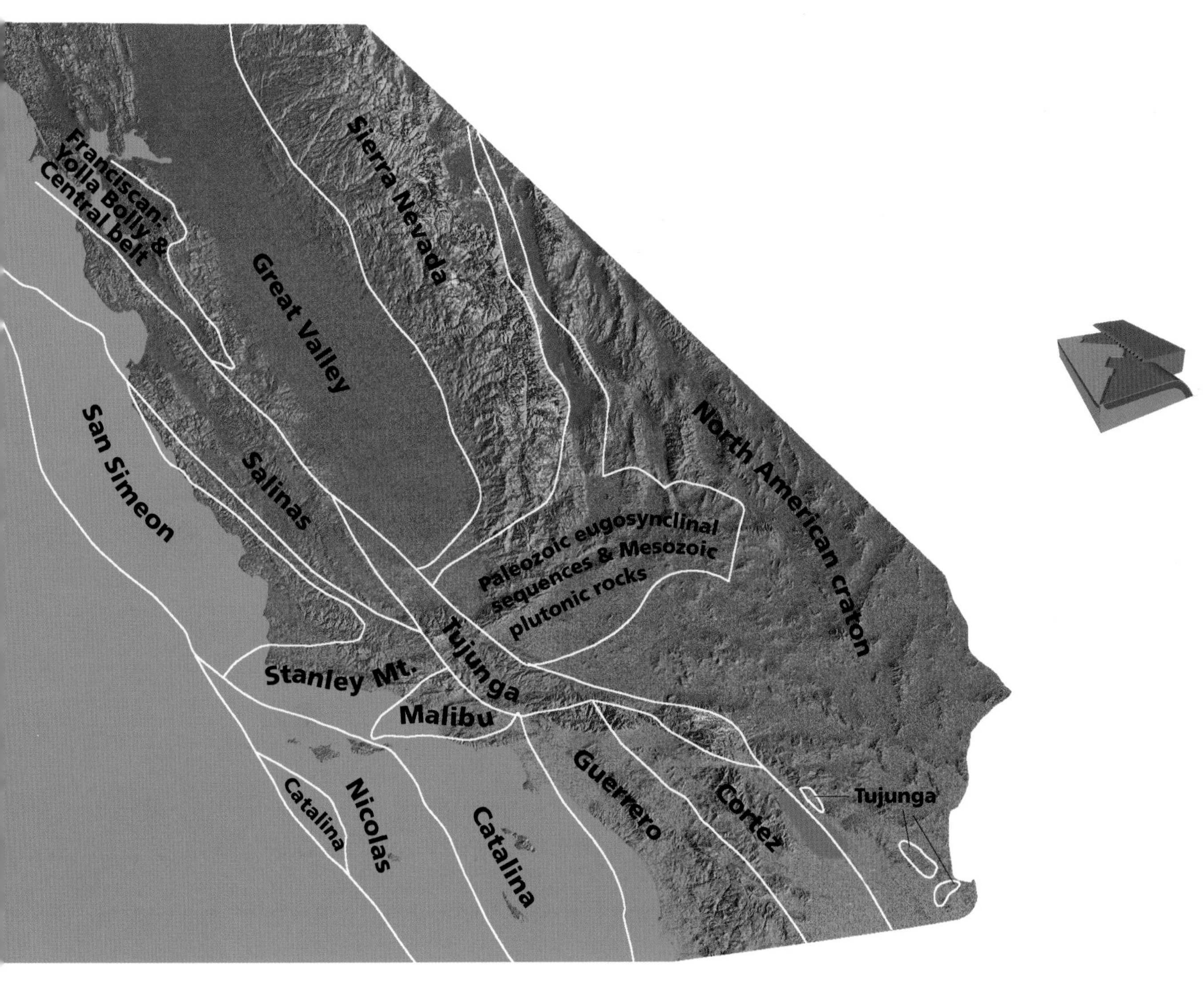

With the steady disappearance of the East Pacific Rise, the strike-slip line of contact between the Pacific and North American plates became longer and longer; it now stretches fifteen hundred miles, from Point Delgada to Mazatlan. As already mentioned, strike-slip motion did not initially begin along what we know today as the San Andreas Fault. At first, older faults such as the San Gregorio-Hosgri (in central California) and the Clemons Well-Fenner-San Francisquito (in southern California) accommodated much of the slip between the two plates.

Bob Powell, a USGS geologist, has spent twenty-five years studying the San Andreas in southern California. He has reconstructed an intriguing story of activity along the three segments of the Clemons Well-Fenner-San Francisquito Fault that began as early as twenty-two million years ago. Entire blocks of southern California began to rotate, and these early fault segments pivoted out of alignments that could accommodate relative movement of the Pacific and North American plates. By thirteen million years ago, the Clements Well-Fenner-San Francisquito faults had locked up. But the plates continued to move and new faults necessarily broke through the crust to relieve pent-up strain: the Canton Fault (thirteen to ten million years ago), and then the San Gabriel Fault (ten to five million years ago). The San Gabriel Fault is visible today as a great crescent slicing across the northern boundary of the Los Angeles and San Fernando basins.

By five million years ago, the San Andreas had come to absorb the lion's share of intraplate strike-slip motion. In northern California, this migration of strike-slip motion has been steadily toward faults that lie increasing distances inland. Now when the Calaveras and Hayward faults slip, they relieve stresses that would have otherwise been focused on the San Andreas. But in southern California, the strike-slip motion stays roughly in the same place while blocks of the crust rotate. Each new fault in southern California dissects through the old faults as blocks keep turning.

Bill Dickinson is a sedimentary petrologist who has lived at the eye of the plate tectonics storm, first at Stanford and then at the University of Arizona. Based on known plate velocities, he concluded that the total expected displacement between the North American and the Pacific Plate should have been 458 miles over the last sixteen million years (when the San Andreas first began to

Farming along the San Andreas Fault at Bitterwater

move). But after a full century of study, geologists have only been able to document 195 miles of slip on the fault. Not all intraplate motion must occur on faults of the San Andreas system, however. Bernard Minster and Thomas Jordan showed how extension of the Basin and Range Province contributes 137 miles to the overall northwest translation of coastal California. This left geologists looking around for an additional mechanism by which to move the California coastline another 126 miles closer to Alaska.

By looking at magnetic fingerprints originally frozen into basalt (i.e., which way was north ten or fifteen million years ago), geophysicists have tracked rotations of rocks in southern California. Dickinson realized that these blocks were rotated by the same intraplate forces that cause right-lateral movement on the San Andreas Fault. As the blocks pinwheeled clockwise in place, their interactions with each other resulted in left-lateral movement across the Garlock and other "backward" strike-slip faults.

After all is said and done, the coherent rotation of large blocks of southern California adds significantly to the overall northwest displacement of coastal California. With this rotation, Dickinson is able to geometrically account for an additional 129 miles of northwest motion. When added to the well-established amounts of transform fault slip (195 miles) and Basin and Range extension (137 miles), the sum is 461 miles. This is within three miles of the total movement predicted to have taken place between the Pacific and North American plates over the last sixteen million years. Not bad for government-funded work.

Oak in Cholame River Valley near Parkfield

WHAT CAN WE LEARN about the earth's interior by studying the San Andreas? Geologists love to wave their arms wildly like crazed symphony conductors. They slash open faults with the enthusiasm of a violinist playing the "Flight of the Bumblebee"; they crash continents into one another like cymbals, and blast open volcanoes as though they were trumpets. But it is also possible to hear other more subtle notes while exploring the San Andreas.

A flying rock once cracked my truck's windshield. I watched first with dismay (before remembering that I had insurance) and then with growing fascination as a crack propagated out from the little impact crater. The crack had a mind of its own, nosing this way, then that, over the next few days. When sunlight warmed the glass, or when the truck rattled down a bumpy road, the windshield was subject to higher stress and the crack spread faster. When I stopped, the crack slowed down. Fractures propagating through rocks on the San Andreas are fundamentally similar. As underlying plate motion increases stress across the fault, fractures are more and more likely to be initiated and to spread. Little earthquakes frequently happen along the fault—a rock popping apart here, a boulder splitting there. But the fault's surface varies from soft to hard material. Small fractures sometimes cease to propagate when they encounter either a body of hard rock that refuses to budge (this time), or a bed of loose sand that inadequately transmits the sudden shock of fracture.

Mark Zoback is a Stanford geologist. How strong, he asks, is the San Andreas? Does it forcefully resist slipping, or is it a weak zone that moves at the drop of a continent? The answer to this question has obvious consequences: if the fault is very weak, the opposing edges can't build up much strain before movement occurs, and therefore earthquakes might not be as damaging. Furthermore, preferring this weak zone, earthquakes should occur over and over again right along the fault rather than migrating to other sites with more resistance to slippage.

In the 1980s, Zoback oversaw the drilling of a two-mile-deep hole on the San Andreas Fault at Cajon Pass north of San Bernardino. He measured heat flow within the hole, figuring that heat generated by friction would be greater if the fault was "strong" and could effectively resist slipping. He also observed deformation of the walls of the hole, trying to understand the direction in which pressures were being applied across the San Andreas. If a great deal of compression held the opposing sides of the fault tightly against

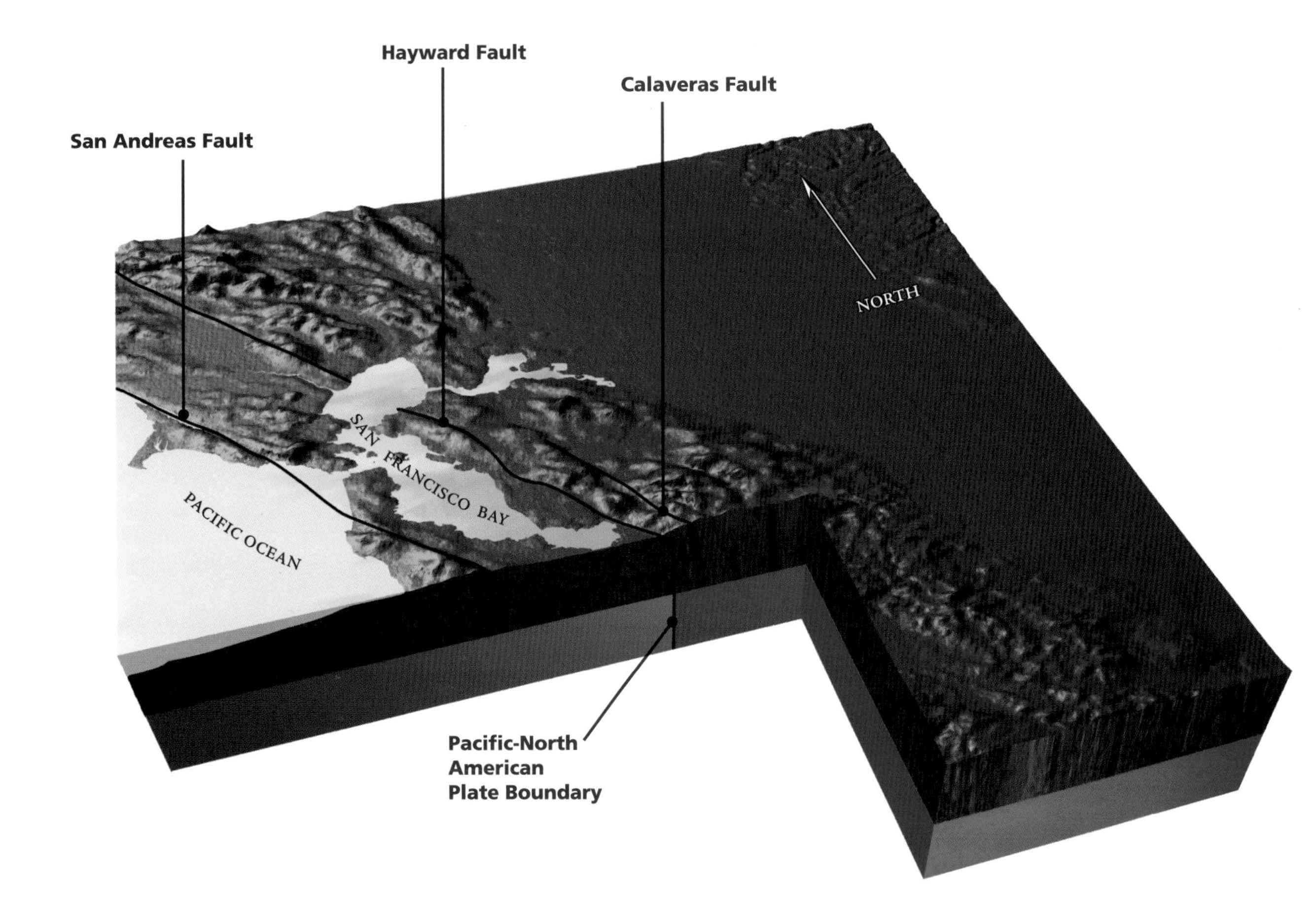

The San Andreas Fault is considered the boundary between the Pacific and North American plates, but in fact the boundary may be a composite of a number of faults, including the Hayward and Calaveras. If this is true, then the faults may be connected above the mantle by a horizontal detachment surface.

one another, more force would be required to propagate a fracture. Zoback learned that even though the greatest regional compression is aligned directly across the San Andreas, the fault still moves easily and generates surprisingly little heat. He concluded that the fault is therefore "weak," making it much more susceptible to movement than other faults that might be nearby.

Zoback was able to directly measure physical properties of only the upper 20 percent of the San Andreas Fault. Seismograph tracings of earthquakes suggest that the fault typically extends ten miles down before petering out in the lower ductile portion of the crust—the part that can be molded or shaped. Below this, the crust is too hot and soft, the "rocks" too viscous, to support shear stresses that must build up before they suddenly snap. The full crust extends another five or ten miles beneath the San Andreas to the Moho, which separates crust (above) from mantle (below).

BACK WHEN HARRY HESS was first cooking up the rudiments of plate tectonics, most scientists assumed that oceanic plates were pushed across the face of the earth by the upwelling of mantle plumes at spreading centers. But more recently, some geophysicists have changed their minds and decided that oceanic crust is pulled by its leading edge, which descends into a subduction zone. Pushed. Pulled. So what? If oceanic crust is indeed pulled down into the mantle, then a significant gap between the Pacific and Farallon plates could have appeared as the East Pacific Rise descended into the subduction zone beneath California. When this happened, gaps or slabless windows opened beneath North America, and hot mantle magmas welled up against the underside of the continental crust. Beginning twenty-four million years ago, three spasms of volcanism occurred within central and southern California, each timed to subduction of different parts of the East Pacific Rise and the consequent development of slabless windows.

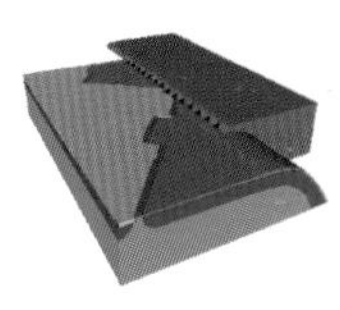

Over the years, seismologists have assembled huge data files from thousands of earthquakes, and from them, are beginning to assemble a three-dimensional image of the earth's deeper mantle, much as physicians use a CT-Scanner to "see" inside a human body. Slowly, a picture is swimming into focus of ancient foundered ocean crusts hundreds of miles within the earth. Gradually, we are learning to move real slow and "hear the rocks talkin'."

In the Field: Bob Powell

ON HIS THIRD DAY as a Cal Tech graduate student, Bob Powell had an epiphany that would forever change his life. The southern California smog had finally parted, revealing the San Gabriel Mountains towering above Pasadena. Powell had moved here from the East after a stint in the navy—experiences in a flatter world had not prepared him for California's vertical views. He was stunned. From this point on, these mountains would dominate his professional career as a geologist with the USGS.

Powell has spent the last twenty-five years mapping the Transverse Ranges of southern California from the San Gabriels to Joshua Tree National Park. He is proud to be, first and foremost, a field geologist, slowly amassing a fundamental familiarity with the rocks of these ranges. On a field trip to the Vasquez Creek Fault, we lurched to a stop at Millard Creek and piled out to examine rocks that had been transported downstream by erosion. Augen gneiss, metasedimentary pelitic schist, Jurassic diorite, and the lovely dappled Mount Lowe granite. Powell was at home among these hard rocks with intimidating names, easily identifying each as though they were old friends. We pondered dacite dikes and horneblende-rich metamorphics on the southwest side of the Vasquez Creek Fault, clearly different than the suite of rocks found on the other side at Millard Creek. Powell grappled with their deep weathering, trying to piece together a credible history of movement on the fault.

The Vasquez Creek Fault appears to connect the old San Gabriel Fault with zones of thrusting that run along the base of the mountains—zones collectively referred to as the Sierra Madre Fault—along which the San Gabriel Mountains have been shoved toward the Los Angeles Basin. I found it fascinating to watch Powell as he tried to sort out cross-cutting relationships between these faults. The game here is to predict which faults are still active and could pose a threat to homes within the developments that creep inexorably up the base of the mountains. Powell was particularly interested in subsidiary high-angle faults, which could act as nucleating surfaces for landslides. We examined traces of the huge Hennigen Flats landslide; movement here had most likely occurred a few hundred thousand years ago during the more moist Ice Ages of the Pleistocene. Powell wondered if Hennigen Flats could be a harbinger of future slides.

Bob Powell is one of many USGS scientists who are trying to apply knowledge of geologic processes to our daily lives. The work is slow, sometimes tedious. But gradually, given persistence and a some measure of luck, geologists like Powell are able to offer a clearer understanding of how the face of the earth evolves.

Bob Powell examines a small fault in the San Gabriel Mountains.

The Face of the Fault

uilla Indians
rtheast shore of
sor of today's Salton Sea.
higher then, and the water
salty. The Cahuilla fished in the
gathered whatever sustenance the
unding desert and foothills would give up.
At the village, they arranged stones in circles
about four feet in diameter. We can only
conjecture about the rings' use; perhaps they
were sleeping circles, or the bases of now-eroded
small shelters, or perhaps even prayer sites.

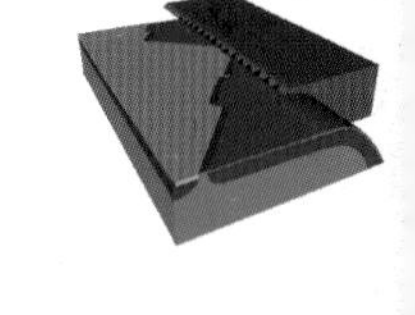

One of the rings lies at the base of a small bluff. The face of the bluff is unusually steep and linear, a scarp created by a fault. Acorn shells are buried near the ring, dated by radiocarbon assay at about eleven hundred years before present. Oaks haven't grown here in many thousands of years; whoever laid these stones also brought in the acorns. Unlike eight or ten neighboring rock circles nearby, this

Santa Rosa Mountains and the Salton Sea

Geologist Howard Shifflett points out the offset stone rings near the

particular ring has been split right down the mi
Andreas Fault. Half of the ring has been carrie
northwest, transforming the once-closed circle i   ıape.
The same strand of the San Andreas has cleave   d
dislocated nearby stream channels.

Why in the world would the Cahuilla India   wanted to build their village so near the San Andreas Faul   'y they had zoning restrictions to prevent such foolishness.   ıaps they didn't realize that a major plate boundary ran right be   :ath their fish-drying racks. Just a few miles to the northwest, Painted Canyon drains out of the Mecca Hills. There can be no mistaking that something dramatic has been happening. The canyon's Palm Spring Formation conglomerates date back only about three or four million years—barely dry behind the ears in geologic terms. Even so, these young rocks have been wildly tipped this way and that by movement associated with the San Andreas. The Mecca Hills were pitched skyward while erosion simultaneously sliced deep canyons back down into their flanks. The San Andreas has wreaked havoc here.

There are other stretches of the fault where its basic directions and styles of movement are easier to envision and understand. The Carrizo Plain is out there—way out there—about midway between San Luis Obispo and Bakersfield. The Carrizo is a perched and

---

Whitewater Canyon in the eastern San Bernardino Mountains; splays of the San Andreas cut diagonally across the canyon.

White evaporites (sedimentary deposits formed by precipitation from evaporating seawater) line the bottom of Soda Lake on the Carrizo Plain

parched basin, two thousand feet above sea level but without drainage to the outside. The plain receives only eight and a half inches of rain a year. Soda Lake, at the north end of the basin, collects what little moisture happens to run down from the surrounding treeless hills. Water that reaches the lake can escape only by evaporation, and in the process, leaves behind its load of sodium sulfate, sodium chloride, and other salts. This may sound like forbidding country, but to the giant kangaroo rats and the blunt-nosed leopard lizards, it's a little slice of heaven. In country this dry, erosion moves at a snail's pace, unable to keep up with or obscure the movement of the San Andreas Fault.

The Temblor Range bounds the Carrizo Plain to the northeast, rising a couple of thousand feet above Soda Lake. Washes that are normally dry stretch down the range's southwestern side toward the plain. Washes carve arroyos on the higher slopes, redepositing sediment farther down the pediment that stretches to Soda Lake. Near the base of the Temblor Range, these normal processes of

The San Andreas carries the Carrizo Plain to the right, as seen here beyond Wallace Creek.

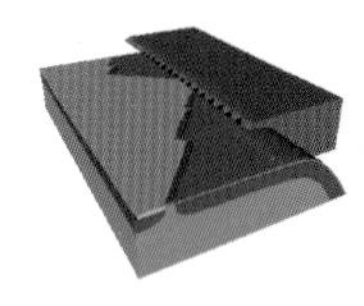

erosion and deposition suddenly go haywire. The washes, apparently in defiance of all rules of gravity and geomorphology, suddenly jog first right, then left—and then continue downslope as if nothing had happened. These jogs accurately trace the course of the San Andreas Fault. The fault in this vicinity moves fitfully, jumping sideways in increments of ten or twenty feet every 150 years. The downstream ends of the channels are carried along with the Pacific Plate as it moves northwest relative to the rest of the Temblor Range and North America.

Of course, the work of gravity is not denied here, just delayed. When intermittent floods roll down these offset stream channels, they initially find it easier to follow a longer course through the broken ground of the fault than to cut new straight channels. But the longer course creates a lower gradient. Sediment carried by a flood drops out of suspension just above the jog and begins to fill the channel; each subsequent flood has a harder time maintaining its momentum. Eventually, the stream abandons its crooked course and

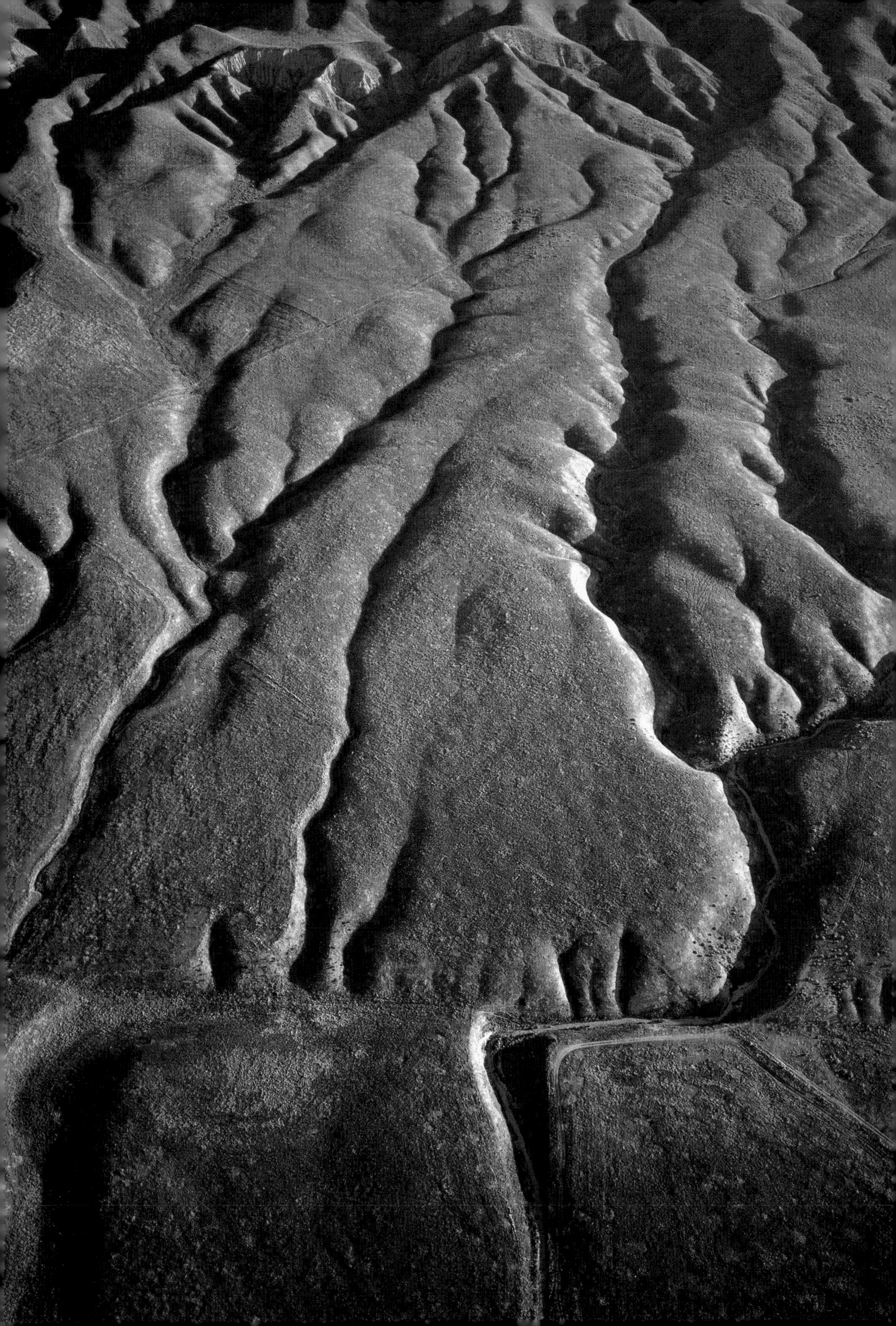

once again flows straight downhill. Until, that is, the San Andreas throws another jog into its course, such as the offset created by the area's most recent great earthquake, which was centered at Fort Tejon in 1857.

Stand on either side of the San Andreas Fault and look across. Land on the other side has moved to the right. This "offset to the right" is true of the entire San Andreas throughout its 750-mile course through California. Thus its descriptive name as a *right-lateral strike-slip* fault. In the Carrizo Plain, as elsewhere along the San Andreas, horizontal movement outweighs vertical movement by a factor of ten or twenty to one. Some vertical movement does take place, however. That is how the Temblor Range came to stand guard above the Carrizo Plain. Small elongate hills called *shutter ridges* lie parallel to the fault, the result of very local compression and uplift adjacent to the fault.

Robert Wallace of the US Geological Survey described these ridges and other fault features within the Carrizo Plain in 1968. Twenty years earlier, he began a remarkable career in geology by studying the San Andreas further to the south near Palmdale; by starlight he ate cold beef stew from a can and serenaded the coyotes with his violin. As a graduate student in 1949, he went out on a limb and suggested that the San Andreas Fault near Palmdale might have slipped seventy-five miles—an unimaginable amount to scientists of the time. Most people agree today that the total slip on the fault is four times this figure, but in its day, his was a shocking hypothesis.

Wallace carefully mapped faults within the Carrizo Plain and showed that here, the San Andreas is actually a fault zone made up of multiple fault strands oriented en-echelon—each strand moves sympathetically with the others in a right-lateral slip sense. Wallace's work was extended by Kerry Sieh, a Stanford graduate student in the late 1970s, who painstakingly mapped out a composite history of recurrent faulting along the San Andreas in the Carrizo Plain over the last thirteen thousand years. Sieh was able to show that major earthquakes strike here on average every century and a half. Fort Tejon in 1857. Once every 150 years. Guess what...

The Carrizo Plain is at the southern end of the long, more-or-less straight portion of the San Andreas that stretches south from

---

Wallace Creek streambed, offset by the San Andreas Fault

Destruction in downtown San Francisco after the April 1906 earthquake

Looking northward on Howard Street at 17th Street, San Francisco, 1906. The rail at right foreground settled and moved forward. The buckle of railway tracks resulted from this movement.

Bailey's pier, northwest shore of Tomales Bay, 1906. Previous to the earthquake, this pier was straight. Its present form is due partly to restoration, but chiefly to the shifting of the muddy bottom of the bay.

the Mendocino coastline. Farther south, the fault bends about thirty-five degrees, first through a left and then a right kink, to resume its generally northwest-southeast trend on the way to the Salton Sea. Within the straight portion to the north, there are a number of elongate valleys that follow the fault's path. San Andreas Lake and Crystal Springs Reservoir are south of San Francisco. Tomales Bay is north of the Golden Gate in Marin County, cleanly separating Point Reyes from the rest of California.

The Plantation House is still farther north, in the hills above Fort Ross. A stage stop when first built in 1871, the house sits in the middle of six parallel fault strands; by chance it happens not to be directly above any one strand. All six broke through to the surface during the great earthquake of 1906, but the house survived. A few feet away, a huge redwood tree was torn in half, its northeast side attached to the North American Plate, its southwest side traveling a foot to the northwest with the Pacific Plate. (The tree was able to heal this wound and continues to grow today.)

One expects to find dramatic evidence of a geologic feature as immense as the San Andreas Fault. Shattered redwoods, great lakes

Fence offset by 8½ feet after 1906 earthquake

Blue stakes on the hillside trace the path of the San Andreas Fault as it runs under W. D. Skinner's barn, now part of Point Reyes National Seashore.

and long lagoons, mountains hurled into the sky—these should be the norm. But they aren't. In many places, indeed most places, you have to look hard to find the fault. An easier way is to follow the signs to the Earthquake Trail at Point Reyes. In less than a mile, this trail loops out to the San Andreas and back, highlighting the scarps and offsets caused by the great San Francisco earthquake of 1906. The trail approaches the farm that once belonged to W. D. Skinner.

Mr. Skinner certainly found dramatic evidence of faulting when he awoke on April 18, 1906. The eastern corner of his barn had been torn off and flung to the southeast. His neatly aligned raspberry bushes were suddenly offset by fourteen feet. Once-straight fence lines across the hill behind his house abruptly jogged fifteen feet.
G. K. Gilbert—then and still, one of history's preeminent geologists—paid Mr. Skinner a visit soon after the earthquake. Gilbert reported that Skinner's most sobering memento was the tail of a cow sticking up from a trench that had supposedly opened and then closed during

San Andreas-created scarps, seen here above a sag pond near Parkfield; earthquake-created scarps are often sharp, but in a damp climate, erosion quickly wears them down.

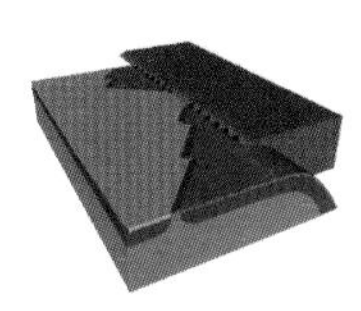

the earthquake. Subsequent reports suggest that this may have been the work not of God but of a quick-thinking farmer who wanted to dispose of an ailing cow. Gilbert did a better job of documenting three phases of deformation during the quake: ridge, trench, and en-echelon. These correspond to local zones of compression, tension, or pure shear across the fault during the earthquake.

Northern California's coast can be damp; thirty inches of rain fall on Point Reyes Station each year. Time has softened the scarps that run through Skinner's ranch. The surrounding hills curve and roll because the processes of erosion are accelerated by all that moisture. Right on the San Andreas Fault, the landscape is held in a delicate balance between the tectonic jostling of the fault and the smoothing hand of erosion. The ragged surface breaks that Gilbert documented ninety years ago have been air-brushed into contoured hills. Here and there, a steeper-than-expected scarp faces northwest or southeast. Sometimes a geologist must back up and squint in order to see the world more clearly.

These oaks within the Los Trancos Open Space were knocked down in 1906, but immediately resprouted. Tree-ring core dating reveals that the horizontal limb is older than 1906 and the vertical limbs are younger.

The reconstructed fence in the hills above Palo Alto shows offsets typical of the 1906 earthquake.

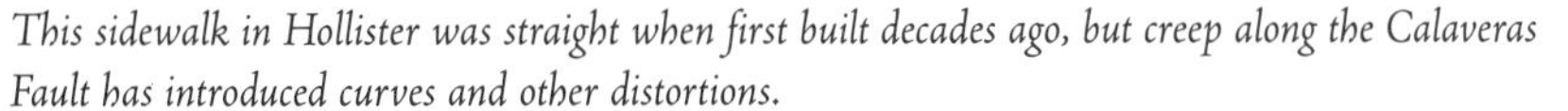

This sidewalk in Hollister was straight when first built decades ago, but creep along the Calaveras Fault has introduced curves and other distortions.

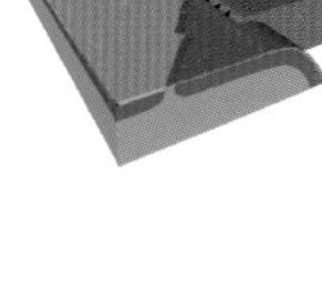

The Los Trancos Open Space Preserve above Palo Alto highlights another stretch of the San Andreas that was torn by the 1906 earthquake. Its Fault Trail starts at Page Mill Road and explores the headwaters of Stevens Creek. The creek heads here because over time, rocks crushed by the fault have been easier targets of erosion than the surrounding unfaulted rocks. Down the trail lies an oak tree that was knocked over by an earthquake. Limbs thrown to the ground resprouted; tree-ring cores date their earliest regrowth to 1906. Offset fences record only three feet of movement, far less than at Point Reyes. So far, there have been no reports of cow tails sticking out of the ground.

WHEN GEOLOGISTS think of crustal plates, they tend to see a smooth movie of steady plate motion: California coasting serenely toward Alaska to be eventually subducted within the Aleutian trench. For the most part, this movie is shot at one frame every few decades. Seen frame-by-frame, the motion is jerky. 1857. 1906. But when the film runs through a geologist's projection of time, it smoothes out. The music comes up. And California sails off into the sunset.

There are places along the San Andreas, however, where the movie advances by one frame every few weeks. Hollister is a town that straddles the Calaveras Fault, a subsidiary branch of the San Andreas. Fault motion here occurs by creep rather than episodic earthquakes. Surprisingly, the San Andreas is one of very few faults in the world that exhibits creeping motion. A homeowner, looking down at his offset sidewalk, once told me that the plates were slowing down; he had only seen an inch of movement the previous year. The creeping segment of the San Andreas Fault stretches ninety miles from San Juan Bautista (near Hollister) southeast to Cholame. Relatively few large earthquakes have been recorded here; instead, plate motion occurs as slow, steady slip.

As seen at Stevens Creek, rocks ground up along a fault are more susceptible to surface erosion. Faults also control how water flows

Cahuilla Indians once took refuge at Thousand Palms, north of Indio. This oasis straddles the Mission Creek Fault, which forces groundwater to the surface at the point at which it encounters the fault.

Sag ponds and the San Andreas along Mustang Ridge

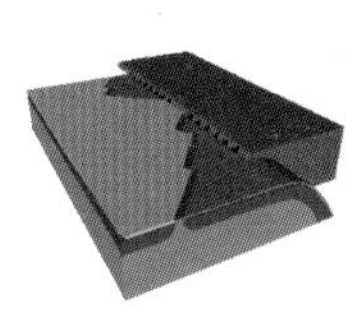

underground. All else being equal, water will move more freely through rocks that have been broken. Sometimes the fault will counterpose a porous rock such as sandstone against another rock—granite perhaps—through which water flows less freely. When groundwater encounters the San Andreas in these places, springs pop to the surface. Palm oases dot the fault near Indio along small seeps that make life possible in an otherwise barren desert.

When ground up along a fault, some rocks are pulverized to fine clay. This clay can create an impermeable barrier to groundwater. If a fault has dropped a block of earth downward and created a basin, any water that it traps may not be able to drain away because of clay lining the pond. The resulting basins become sag ponds, ubiquitous tiny lakes found along the San Andreas.

If you're ever driving from San Lucas to Coalinga on California Highway 198, keep an eye out for the San Andreas Fault. Cresting Mustang Ridge, the roadbed is broken, its foundation settled. Don't be surprised to find an orange CalTrans truck parked on the shoulder—always a good clue that the San Andreas is nearby. But

Marine terrace, beveled flat by waves, now lifted above sea level south of Morro Bay

you'll encounter an even better clue a few hundred yards to the east: a straight line of ponds stretching both northwest and southeast away from the road. These sag ponds sit awkwardly atop the ridge. Any "natural" body of water would have drained off this ridge eons ago. But the San Andreas has been able to drop the ponds' bottoms faster than sedimentation can fill them in again, faster than erosion can chew away the sides.

The fault's trace can also be found within other larger bodies of water, such as the Pacific Ocean. Sonar mapping of the ocean floor between Point Arena and Shelter Cove traces the San Andreas as it runs seventy-five miles parallel to and just offshore of the Mendocino coastline. Shelter Cove is the last place the fault touches dry land before swinging west toward its northern terminus at the Mendocino Triple Junction. Mapped underwater, the fault is revealed as a series of west-facing underwater scarps.

Noyo Submarine Canyon extends out from the coast but is deflected to the right as it crosses the San Andreas, just like those stream channels on the Carrizo Plain.

The coastline from San Francisco to Monterey offers another more uplifting vision of the San Andreas. The San Andreas itself lies ten or fifteen miles inland, within or just northeast of the Santa Cruz Mountains. Highway 1 is perched on a series of flat benches that hover above the ocean. Each bench was at one time submerged just below the ocean's surface, ground flat by waves, and then uplifted as much as 400 to 600 feet over the last million years. Each successively higher bench records a previous cycle of uplift and seafloor erosion. Similar benches are found south of Point Arena, at San Simeon, and near Morro Bay.

South of Monterey, Highway 1 clings to cliffs that have obviously jumped right up out of the ocean. The interaction of the Pacific and North American plates has produced disruptions immediately along the San Andreas Fault as well as at distances of many miles. Much of the California coastline is being uplifted—very quickly in geologic terms—because of the interactions of the two plates. Coastal Californians would do well to worry less about falling into the ocean and more about being lifted so high that they would have to trade their surf boards for snow boards. If Santa Cruz continues to rise at current rates (and if the processes of erosion are simultaneously suspended), its famous boardwalk will be almost seven thousand feet above sea level within the next five million years.

THE SAN ANDREAS SEEMS so impossibly enormous. We Lilliputians look for the tracks of its motion that might be meaningful in our own small framework of space and time: scarps, sag ponds, offset streams, uplifted terraces. But perhaps my favorite is the local absence of balanced boulders. James Brune, who teaches at the University of Nevada in Reno, wandered up and down California looking for precariously perched boulders. He found none in areas along the fault that had been shaken by strong earthquakes. Sometimes geology is wonderfully simple.

In the Field: Michael Rymer

MICHAEL RYMER WAS in the vegetable section of a Coachella grocery store when the 1992 Joshua Tree earthquake struck. The ceiling panels fell out but the bell peppers stayed put. Rymer bought a flashlight and raced to the San Andreas Fault. Lucy Jones, his USGS colleague in Pasadena, told him by phone the coordinates and focal plane solution of the earthquake. Thirty minutes after feeling the shock waves, Rymer suggested to Jones that the quake had occurred along faults first mapped by Rymer only months before the Joshua Tree earthquake.

Statistically, Michael Rymer is not the safest geologist to accompany in the field. Between growing up in the Bay Area and his twenty-five years with the Survey, he has experienced hundreds (if not thousands) of earthquakes; seven were big enough to be named. He is a veritable earthquake lightning rod. Nonetheless, I joined him one cool spring day, exploring the deep canyons of the Mecca Hills. Scrambling up the steep gullies was difficult—my fingers were constantly crossed and I was hoping that a major earthquake would not roll through while we were in those canyons.

Rymer has studied the San Andreas Fault in the moonscape deserts near Indio and the Coachella Valley since 1985. Where else, he asks, is the fault so perfectly exposed? He has mapped the Palm Spring, Ocotillo, and Imperial formations in excruciating detail; found fossil skulls of Pleistocene horses and cotton rats; chased 760,000-year-old Bishop Ash from one side of the valley to the other; pinpointed magnetic reversals within sandstones and mudstones; and plotted the rotation of magnetic vectors of rocks

The San Andreas Fault has dramatically tilted the sedimentary Palm Spring Formation from its original horizontal position to the nearly vertical orientation found in the Mecca Hills.

Michael Rymer is guilty here of that most egregious of geologic sins—arm-waving—above the Painted Canyon Fault in the Mecca Hills.

spinning between the Mission Creek and Banning faults. He has painstakingly gathered these bits of the San Andreas puzzle over thirteen years of research. Each piece illuminates the others.

A hiker once asked Rymer what he was doing in the Mecca Hills—looking for gold? No, he said, looking for processes. This answer was, of course, met with a blank stare. Rymer has explored these rocks in minute detail and can confidently describe their involvement with the San Andreas Fault. The rocks were deposited as the Coachella Valley sagged and moved to the northwest; they buckled upward as the fault moved. Rymer works not just in the two dimensions of deposition, not just in the three dimensions of folding, but in the four dimensions of rocks folding and faulting through time. He doesn't need to find gold to justify his interests—the ongoing processes of discovery and understanding are their own reward.

From San Bernardino southward, the San Andreas has not experienced a very large earthquake in at least three hundred years.

Some geologists feel that the distribution of smaller, more recent, earthquakes suggests that a very large earthquake on this segment of the San Andreas would originate near the Salton Sea and propagate northwest. But the greatest number of people live in San Bernardino, not at the southeast end where a large quake is perhaps more likely to start. So there is much interest in (and therefore, funding for) analyzing the fault where the concentration of people is greatest. Rymer suggests, however, that this is like studying a gun by examining only the last inch of its barrel.

Swifts dive-bombed us as we reached a false summit in the Mecca Hills. The day was clear, the sky, deep blue. Rymer pointed out a thousand exposures he had mapped, and hundreds of details of the fault he had pondered. He tossed me a rock with a smaller pebble locked inside; he talked about how the Palm Spring Formation was made of reworked sediments from the nearby Little San Bernardino, Santa Rosa, and Orocopia Mountains. I watched him make maps in the air with his hands and knew that he was right: we need to know all aspects of a landscape—from the smallest pebbles to the largest faults—if we are going to understand its behavior.

Whole Lot of Shakin' Goin' On

TO EXPERIENCE AN EARTHQUAKE is to be out of control. Earthquakes: unpredictable, unnerving, and—if you live along the San Andreas Fault—inevitable. You cannot live in California without looking over your shoulder. Let's face it, if the San Andreas did not periodically lurch to and fro, you would not be reading this book. Crustal plates may move all they want, but we tend not to be overly concerned until the movement touches our own lives. What, then, is an earthquake? When, where, and why does it occur?

The Pacific Plate is moving northwest (thirty-six degrees west of north, to be exact) at 1.9 inches per year relative to the North American Plate. Some of this movement is accommodated by the compression and crumpling of parts of California—the San Gabriel Mountains and the Kettleman Hills—and some by the taffy-like extension of large parts of eastern California, Nevada, and western

Shutter ridges along the San Andreas, Carrizo Plain

Utah. But most occurs as slip across the San Andreas Fault system. Kerry Sieh, working on the Carrizo Plain, calculated 1.4 inches of slip on the fault annually when averaged over the past thirteen thousand years.

For a moment, let's imagine that slip across the San Andreas is temporarily arrested but that the Pacific and North American plates are still inexorably moving northwest/southeast relative to one another. In other words, the plates still move but their edges are locked together. The cold, brittle upper crust bends as plate motion continues, bends but doesn't yet break. The strained crust is increasingly charged with potential energy, stored but not yet released. If any material is bent slowly over millions of years, it will eventually assume the new shape into which it has been pushed. If that material is heated—as happens to rocks deep within the earth—it will relax into its new shape a bit more quickly. Given sufficient time and temperature, whole mountain ranges can be stirred and swirled into an amazing array of shapes. But if shallow (i.e., cold and brittle) crustal rocks are bent too quickly (at a shearing rate of, say, 1.4 inches per year), they will eventually break, and rocks on either side will snap back to their unbent positions. Voila: an earthquake along the now-reactivated San Andreas Fault.

As two plates slide past each other, their edges tend to drag. With time, the edges bend, elastically storing up potential energy, until the rock is strained beyond its breaking point. A rupture then occurs, suddenly releasing stored-up energy as an earthquake.

Ever since H. F. Reid proposed this Elastic Rebound Theory after observing the 1906 San Francisco earthquake, scientists have been running around California with various contraptions trying to measure the crust's elastic strain. At first, they used simple optical surveying equipment to measure changes in distance and direction between points on opposite sides of the San Andreas. These techniques were refined with the expensive addition of Very Long Baseline Interferometry (VLBI) and Satellite Laser Ranging (SLR). The Global Positioning System (GPS), which relies on U. S. military satellites, came into its own in the 1990s. Now geologists can inexpensively calculate a station's position and movement with accuracies measured in small fractions of an inch. With these tools, we can see strain building up in the crust before it is released as an earthquake. Perhaps someday this information can be used to help to predict earthquakes.

When rocks do snap, their pent-up potential energy is suddenly released into the surrounding earth. The energy radiates out in two basic ways. "P," or *primary*, waves travel fastest. Similar to sound waves, these P-waves are due to compression of rock particles, with each particle moving forward and backward in a straight line from the origin of disturbance. "S," or *secondary*, waves are due to a shearing or transverse motion that moves rock particles not forward and backward, but side to side or up and down, the way ripples radiate out when a rock breaks the surface of a calm pond.

P-waves race through the earth at speeds of about 4 miles per second within the crust and 5 miles per second in the upper mantle. Because S-waves, which travel at speeds of 2.2 miles per second in the crust and 2.8 miles per second in the upper mantle, are due to shearing, they can propagate only through solids. Since the core is liquid, it does not transmit S-waves. This difference is perhaps best illustrated by remembering your last belly-flop from the high dive: the water (obviously a liquid) transmitted P-wave compression forces, but it could not resist shear as you sank to the bottom. If, in your diving splendor, you missed the water altogether and hit the deck (even more obviously a solid), that material transmitted both P- (compression) and S- (shear) waves.

P-waves are the first waves to reach an earthquake observer, sometimes as an audible thud. This wake-up call will be followed by S-waves that pitch the ground into a rolling motion that is much more destructive to buildings than the passage of P-waves. Imagine

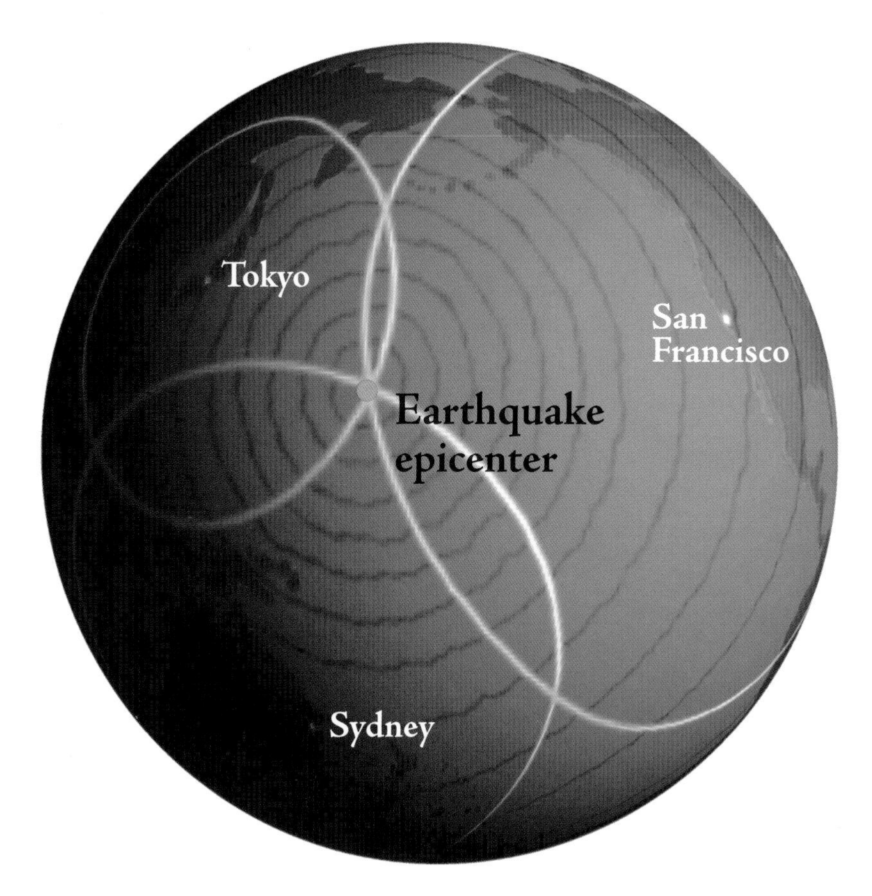

An earthquake's epicenter is located by triangulation, using the distances obtained from three seismic stations.

San Francisco's Transamerica Pyramid as a cork rocking to and fro on a suddenly agitated pond. Since the P- and S-waves travel at different velocities, the interval between their arrivals can be used to calculate the distance from the observer to the waves' site of origin: the longer the interval, the greater the distance.

The Chinese had one of the world's first seismometers: a pot decorated with eight dragons, each with a bronze ball in its mouth. When the vase was shaken by an earthquake, a ball dropped with a loud clang into the mouth of a frog waiting below. The direction from which the earthquake waves had arrived could be deduced by seeing which ball had dropped. California has about 550 seismographic stations, all considerably more modern than the Chinese pot-and-frog models.

A seismograph is a machine that runs continuously and records incoming seismic waves. Today's seismograph is a fundamentally simple two-part contraption: a base with a recording drum attached firmly to the ground and a stylus attached to a suspended weight. Time marks are ticked onto the recording drum at regular intervals. When an earthquake rolls through, the ground rings like a bell as first P- and then S-waves pass underneath the seismograph. The recording drum shakes back and forth with the waves, and the stylus—held still by inertia—records the waves as squiggles on the drum. After an earthquake, a seismologist reads when the first waves arrived at his station and measures the time that separates the faster P- and slower S-waves. If three or more stations scattered around the world record the same earthquake, then the quake's exact point of origin can be pinpointed.

With reports from enough seismographic stations, scientists are able to plot the exact latitude, longitude, and depth of an earthquake's origin beneath the earth's surface. They can show the orientation of the faulted surface: vertical as is most commonly seen along the San Andreas, or close to horizontal as occurs on thrust faults in southern California. By examining an earthquake's very first impulse recorded on the seismograph, scientists know if rocks were thrown toward or away from the station and thus are able to show that a strike-slip fault's movement was either right- or left-lateral.

So how *much* did the earth move last night? We're all familiar enough with the Richter Scale to "Ooooooh" when we hear about a Richter 5.5 earthquake, and "Aaaaaah" when the newscaster mentions 6.5. What do these numbers really mean? Charles F. Richter's scale simply measures the amplitude of maximum seismographic deflection during an earthquake, normalized to a standard sixty-two-mile (or one hundred kilometer) distance between the earthquake and the seismograph. This system tends to misrepresent large earthquakes, however, because near an earthquake's origin, seismograph recording pens can be disproportionately deflected by high-amplitude seismic waves.

The Modified Mercalli Intensity Scale attempts to measure an earthquake in terms of its effects on humans and their dwellings. Intensity I quakes are felt only by the most sensitive observer under perfect circumstances. Intensity VI is felt by every observer, many of whom become frightened. An Intensity XII earthquake causes total destruction.

These scales identify an earthquake's effect either on a seismograph needle or on the population of an entire city. But the system that geophysicists now favor measures earthquake Moment-Magnitude (expressed as M), defined as the energy released by slip on a fault, which takes into account how far the fault slipped, length of fracturing

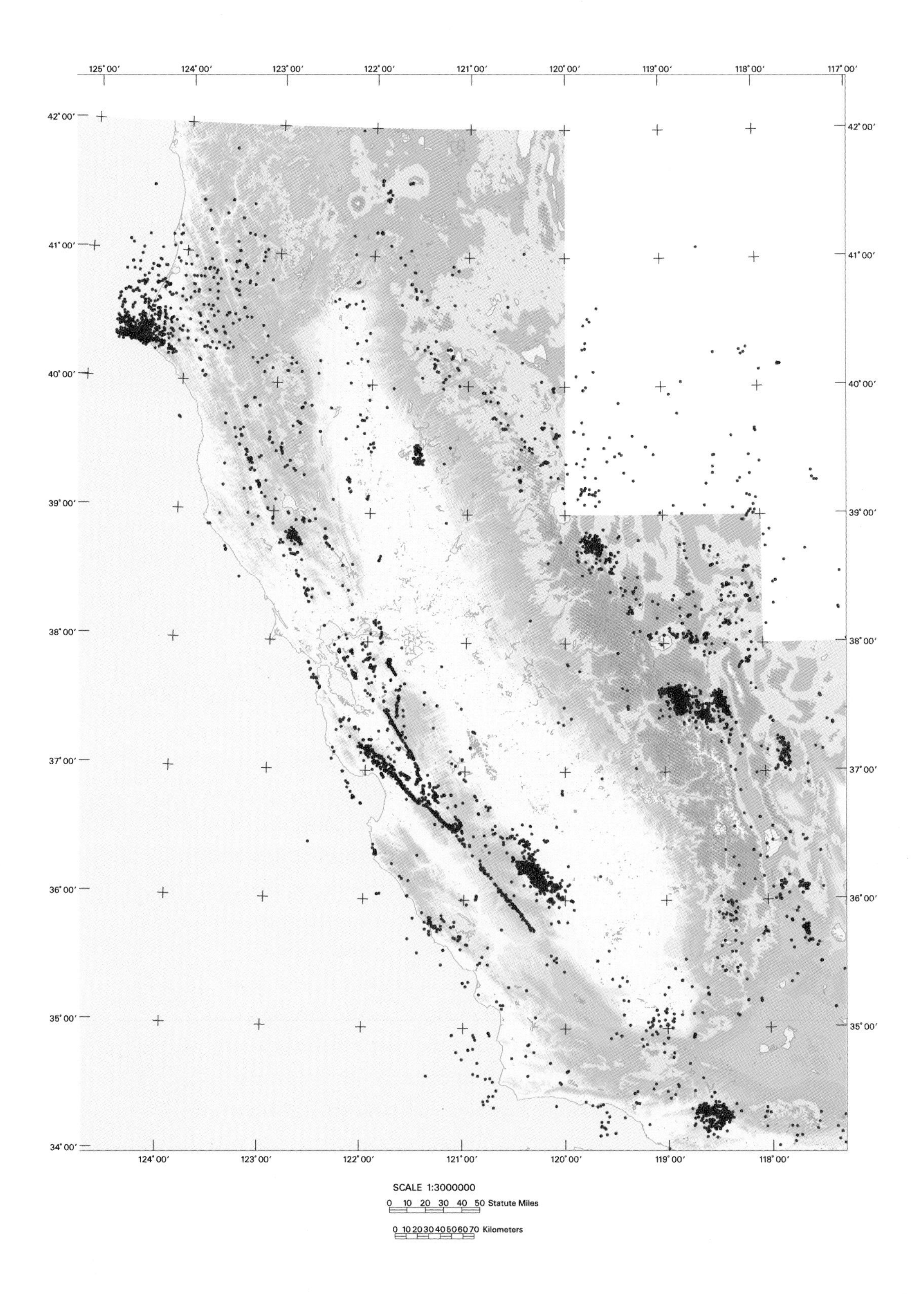

Seismic map of northern California, 1970-1997, showing activity greater than 3.0 magnitude

along the fault, depth at which movement was initiated, and "stickiness" of the surfaces being faulted. During the 1857 Fort Tejon earthquake, at least 225 miles of the San Andreas Fault slipped up to thirty-one feet. The Moment-Magnitude of the 1857 event has been estimated at M=7.8. The 1906 San Francisco earthquake has been assigned a magnitude of M=7.7. Together, these two earthquakes account for fully half of all the seismic energy released along the San Andreas since 1769.

Both the Richter and Moment-Magnitude scales are logarithmic, meaning that each additional unit represents a ten-fold increase in amplitude and a thirty-fold rise in released energy. Even though California is buffeted by many small earthquakes every day, the great earthquakes that roll along once every few decades account for the largest cumulative fraction of movement on the San Andreas and the greatest release of destructive energy.

The *epicenter* of an earthquake lies directly above the point within the earth called the *hypocenter* (or focus), the place where movement starts. Hypocenters along the San Andreas typically are no deeper than 6 to 10 miles below the earth's surface. Presumably, rocks below this depth are hot enough to bend rather than break. Fracturing begins at the hypocenter and spreads upward and outward on the fault plane. This fracture propagation can travel at a velocity of 4,000 miles an hour. The 1906 San Francisco earthquake started just offshore of the Golden Gate; within four minutes, 270 miles of the San Andreas, from Mendocino County to San Juan Bautista, had ruptured.

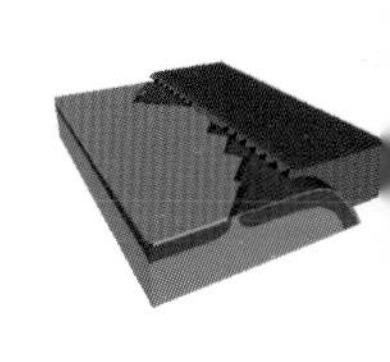

There is a lot to learn by plotting earthquake epicenters on a map of California. Some sections of the San Andreas—around San Juan Bautista, for instance—are tweaked by thousands of tiny quakes. Other places, such as Parkfield, are rocked by modest quakes at regular intervals. But the map reveals an eerie absence of recent movement along the fault north of San Juan Bautista and south of Parkfield. From Shelter Cover to San Juan Bautista, and from Parkfield to Cajon Pass, the San Andreas has been relatively quiet, or locked, since 1906 and 1857, respectively.

The US Geological Survey has examined the San Andreas Fault at Parkfield, where M=6.0 earthquakes (or greater) have occurred on an average of about every twenty-two years since first being observed in 1857. The last one took place in 1966. By 1988, the next quake in the series was overdue. Geologists identified this stretch of the San Andreas as a *seismic gap*, and predicted a moderate earthquake within five years. To date, scientists are still holding their collective breath.

In the Field: Carol Prentice

"I CALL THIS the Blue Goo," Carol Prentice said, patting the dry block of blue-gray mud upon which she sat. "It slid down here sometime in the last thirteen thousand years, based on a radiocarbon date of charcoal in the alluvial layer that's just underneath." We looked past the mud to Shelter Cove, a small hook of land that curls out to sea on the California coast. Shelter Cove harbors what is likely the northernmost exposure of the San Andreas Fault. Perhaps, as Carol suggests, it continues on land a few more miles toward Petrolia. But it's also possible, as other geologists have argued, that instead, the San Andreas lies a few miles west of here, never touching land again after plunging into the sea at Point Arena, sixty miles to the south. Or perhaps, as depicted on most current maps, the San Andreas kisses California goodbye at Shelter Cove, swinging out to sea after passing through cliffs on the other side of this beach.

Carol pulled maps and aerial photos from her backpack, material for a paper on this segment of the San Andreas Fault that she and her co-workers will be publishing in a few months in the Geological

The Lost Coast of northern California, looking toward Shelter Cove

Carol Prentice on the beach at Shelter Cove, and (right) collecting carbon samples for radiometric dating.

Society of America Bulletin. We plotted our position on the beach, and agreed that a fault (presumably THE fault) cut through the hills immediately above us. We found sheared rocks at beach level that certainly showed the trace of some fault.

Carol dug deeper in her backpack, this time coming up with photographs taken by Francois Mathes two weeks after the 1906 San Francisco earthquake and published in Lawson's Carnegie report. Rescued from the depths of Berkeley's Bancroft Library, the ninety-year-old pictures clearly showed fresh scarps. Carol twisted the pictures this way and that, trying to find clues as to which direction the ground had moved: up-and-down or side-by-side? If the motion was predominantly up-and-down, then this was less likely to be the San Andreas. But the aerial photographs clearly showed an offset drainage at Telegraph Creek, about a mile from here, slicing through the Blue Goo. This could only happen if the fault were moving side-by-side. Using her thirteen-thousand-year charcoal date as a benchmark, Carol concluded that the fault at

Shelter Cove is now moving in a strike-slip fashion at a minimum of fourteen millimeters—about half an inch—a year. Too fast to be anything but the San Andreas.

Prentice first studied at Humbolt State University just north of here, preparing to teach general geology at a high-school level. But in her time off, she caught herself reading about faults and earthquakes. This led her, in the 1980s, to work toward her Ph.D. at Cal Tech under Kerry Sieh. Relatively few scientists paid careful attention to the northern end of the San Andreas Fault back then; now, more than a decade later, Carol can still count them on two hands without running out of fingers. Since Cal Tech, she has worked for the USGS as a paleoseismologist, studying geologically recent movements of the San Andreas from the Bay Area northward. And true to her earliest intentions, she still teaches—now, graduate geology students and at seismologic seminars around the world.

Sieh impressed upon Prentice the importance of looking at every shred of available evidence when studying a fault. I watched as Carol passed that knowledge on to Robert Sickler, a student from Humbolt State University. Eight feet down, in a trench across the Maacama Fault at Ukiah, Carol brushed her fingers across sand that felt grittier than the silt in a higher layer. She followed a lens (thick in the middle and thinner toward the edges) of gravel to the point where it pinched out, and then projected that point to the trench's opposite wall. If she could identify a similar line on the other side of the Maacama, she might be able to say something about the fault's magnitude of offset since the gravel was deposited. She carefully traced a recent fracture upward until it vanished; the layer above necessarily post-dates that fracture. Always, she looked for charcoal. Carol Prentice truly knows what it means to work in the trenches.

Pieces of the Puzzle

IT'S TEMPTING TO CONCEIVE of the San Andreas Fault as a single clean boundary between two crustal plates. Would that the world were so simple. The fault is only the most obvious surface along which the plates move. Other faults within the San Andreas system also accommodate plate movement. The plates don't just slide quietly past one another in the night. They collide in some places, separate in others. Compression squeezes flat land up into ridges and shreds the crust along deeply buried thrust faults. Extension draws the earth's crust out like thin dough beneath Death Valley and the Salton Sea. The San Andreas, a complex system, is best considered as a series of segments, each with its peculiar pattern of behavior.

Point Reyes National Seashore (bottom right) to Point Arena (top left), as seen from space

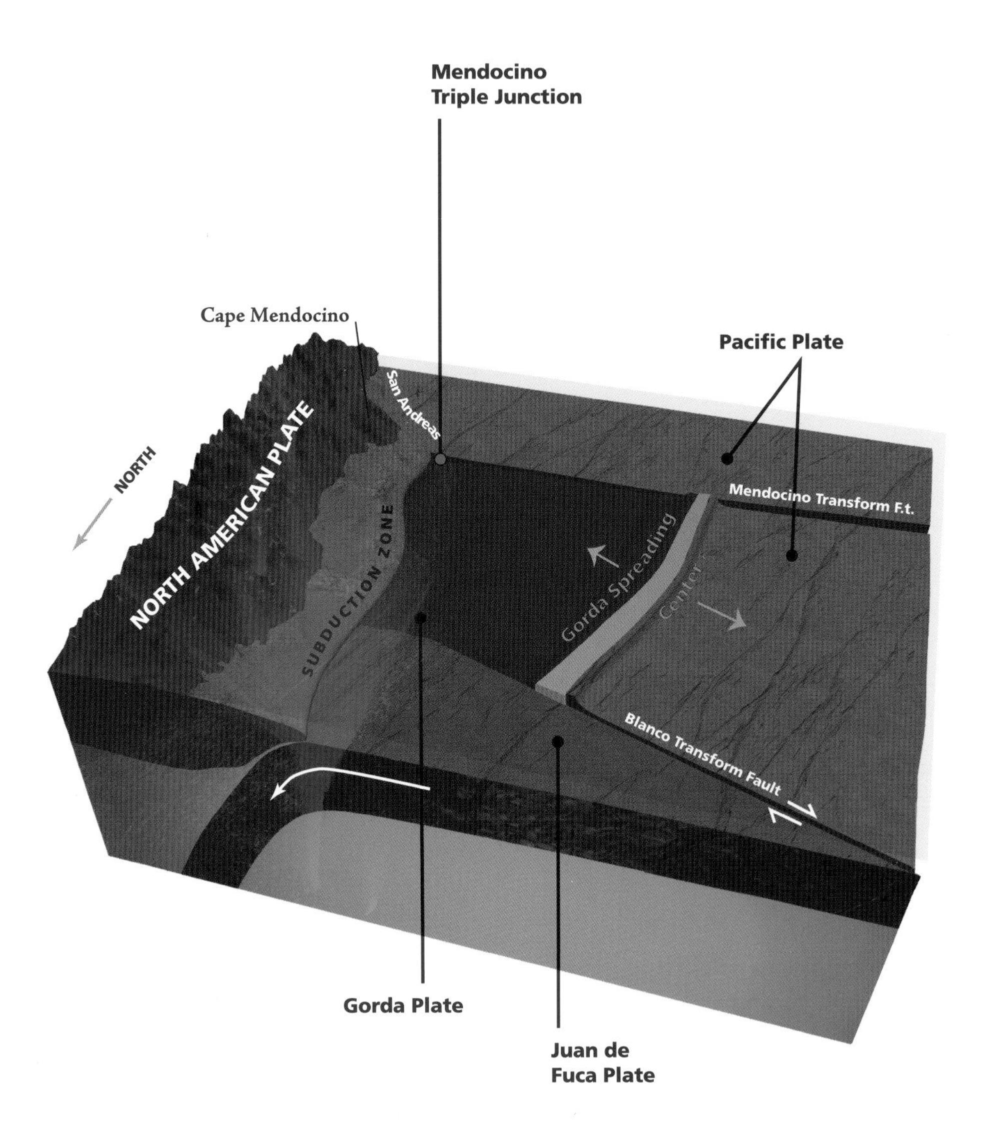

Mendocino Triple Junction

Petrolia is a sleepy northern California town of a hundred people. By and large, life is pretty quiet here. Petrolia is also soberingly close to the northern end of the San Andreas Fault and the Mendocino Triple Junction. This is the point at which three very independent pieces of the earth's crust are in uneasy contact: the Pacific, North American, and Juan de Fuca plates. The southern end of the Juan de Fuca is locally called the Gorda Plate. (Both are remanents of the old Farallon Plate.) These are separated by three major fault zones: the San Andreas (between the Pacific and North American plates) to the south, the Mendocino Transform Fault (between the Pacific and Gorda plates) to the west, and the Cascadia Subduction Zone (between the Gorda and North American plates) to the north.

North of Petrolia and the San Andreas Fault, the Gorda Plate is currently sliding beneath the North American into the Cascadia Subduction Zone. The top of this zone is bounded by a *megathrust* fault that dips beneath the North American Plate, parallel to the Oregon and Washington coastline. This fault seems to slip infrequently, perhaps because it is tightly locked by compression between its upper and lower surfaces. Consequently, more stress has to build up before the fireworks begin. The first documented movement on the megathrust in historic time was recorded on seismographs in 1992. These shocks originated within the Gorda Plate north of the Mendocino Transform Fault, along faults that dipped to the northeast beneath the North American Plate—just like the megathrust. Paleoseismologists such as Brian Atwater of the USGS in Seattle have examined dead stands of drowned western red cedars on the Olympic Peninsula, evidence that five major prehistoric earthquakes have occurred along the Cascadia subduction boundary during the past seventeen hundred years. The intensity of any future earthquake will depend on how much of the megathrust slips. Evidence from prehistoric events points to future magnitudes of at least 8.4 and possibly greater than 9.0 if most of the fault is activated, suggesting that when this megathrust breaks loose, the coasts of Washington, Oregon, and northern California are in for a whale of a ride.

The Mendocino Triple Junction has been rocked by 25 percent of the seismic energy released in California since 1906. When the last seventy-five years of the area's significant epicenters are plotted, most fall within the southwestern toe of the Gorda Plate. By all accounts, this toe is being badly stubbed as it is squeezed by the two neighboring plates. The epicenters document the fragmentation of the plate, its last groan before subduction.

On April 25, 1992, Petrolia's peace was shattered when an M=7.1 earthquake pummeled the town. Houses were pitched from their foundations. Chimneys were tossed about like pick-up sticks. The fire station door jammed, trapping the fire truck inside; the post office, general store, and gas station burned down. Hundreds of landslides blocked roads from Eureka to Point Delgada. Three large landslides above the Eel River delta rushed down the same chutes that slid in 1906. Twelve miles of nearby coastline jumped at least three feet out of the water, instantly stranding mussels and sea urchins above the typical high-tide line. A *tsunami*, a great wave spawned by the quake, swept down the California coast and was recorded four hours later twenty-three hundred miles to the southwest at the Hawaiian Island station of Kahului.

The US Geological Survey and California's Division of Mines and Geology had previously saturated the Petrolia area with a wide array of instruments to measure earthquakes. In addition to regular seismographs, twenty-three strong-motion accelerographs were installed within sixty-eight miles of what turned out to be the 1992 epicenter. Accelerographs are used to record the very large ground motions that occur in the immediate vicinity of an earthquake, motions that would otherwise overwhelm a seismograph of normal sensitivity. One accelerograph at Cape Mendocino recorded a brief vertical ground acceleration more than twice that of gravity. Imagine trying to stand upright if you suddenly weighed five hundred pounds.

The North Coast

Nowhere in California is the concept of the San Andreas Fault *system* better illustrated than along the northern Coast Range from Shelter Cove south to San Juan Bautista. From its vague northern origin somewhere within the Mendocino Triple Junction, the fault is first definitely seen as it briefly makes landfall at Shelter Cove and then travels just offshore to Point Arena. The fault skims behind the Mendocino and Marin coastlines before jumping ship again at Bolinas Bay to cross the Golden Gate. From San Francisco, it runs a relatively straight course down the peninsula toward San Juan Bautista.

The San Francisco Bay Area is sliced into strips by the Hayward, Calaveras, Rogers Creek, Maacama, and Green Valley faults. These begin to splay off the San Andreas near San Juan Bautista, and continue northwest on tracks that lie northeast of and roughly parallel to the San Andreas. They have all demonstrated right-lateral strike-slip motion in historic times. Indeed, these subsidiary faults are responsible for the vast majority of lower-level seismicity along the fault system's coastal section. The San Andreas Fault itself has been as quiet as a church mouse since 1906. But that episode still reverberates in the Bay Area's memory.

At 5:12 a.m. on April 18, 1906, the northern segment of the San Andreas Fault ripped apart from Shelter Cove to San Juan Bautista. The ground convulsed for forty-five to sixty seconds, leveling large parts of San Francisco, San Jose, and Santa Rosa. Witnesses heard an approaching roar and were nauseated by the sea-like undulations of normally solid ground. Horses stampeded out of their traces; cows later had second thoughts about giving milk. Chickens squawked, cats bristled, and dogs jumped out of second-story windows. People eating

The San Francisco Bay Area, as seen from space

Aptos Creek, Forest of Nisene Marks State Park near Santa Cruz

breakfast two hundred miles to the east, in Gardnerville, Nevada, didn't feel the quake but became nauseated anyway.

The M=7.7 shock was felt from southern Oregon to Los Angeles, and inland to central Nevada. San Francisco sustained its greatest destruction in the ensuing fire; residents had broken water lines and dry hoses with which to battle the flames that steadily devoured block after hapless block of homes and businesses. Officially, seven hundred people died; realistically, the death toll was at least three times higher.

The 1906 earthquake began just outside the Golden Gate, where the fault jogs around a small down-dropped basin. The fracture tore northwest past Olema and Point Reyes, where offsets reached twenty-one feet. Ground ruptures could be traced past the Gualala and Garcia rivers, where the fault trace plunges into the sea at Manchester Beach. Andrew Lawson, principal author of the 1908 Carnegie Institution study, found breaks as far northwest as Shelter Cove, leading him to correctly postulate that the fault lay hidden offshore between there and Point Arena. The tear also propagated southeast from its point of origin, but offsets were of smaller magnitude: ten to seventeen feet at Crystal Springs Reservoir and at least three feet in the hills above Palo Alto. (Of possible religious significance were the sulfurous fumes reported to rise from a crack just south of Los Gatos.)

Lawson spent a lot of time twisting his mustache over the paucity of vertical uplift observed on fault traces that ruptured in 1906. To his turn-of-the-century way of thinking, fault motion was expected to lift rocks up or drop rocks down. Perhaps this is a reflection of geology's nineteenth-century mining heritage, the source of terms such as hanging wall and foot wall. I have a clear mental image of Lawson sitting on a rock, grass stem between his teeth, astounded that so much of the observed fault motion was side-to-side. To be sure, he did find some evidence of vertical movement: drowned patches of pickleweed on one side of Bolinas Lagoon, stranded clams on the other. Lawson would probably appreciate the modern geologist's sentiment that the entire San Francisco Bay is a downdropped basin controlled by vertical components of movement on the San Andreas and Hayward faults. But, preconceptions notwithstanding, his overwhelming observation was that the San Andreas had moved primarily side to side rather than up and down.

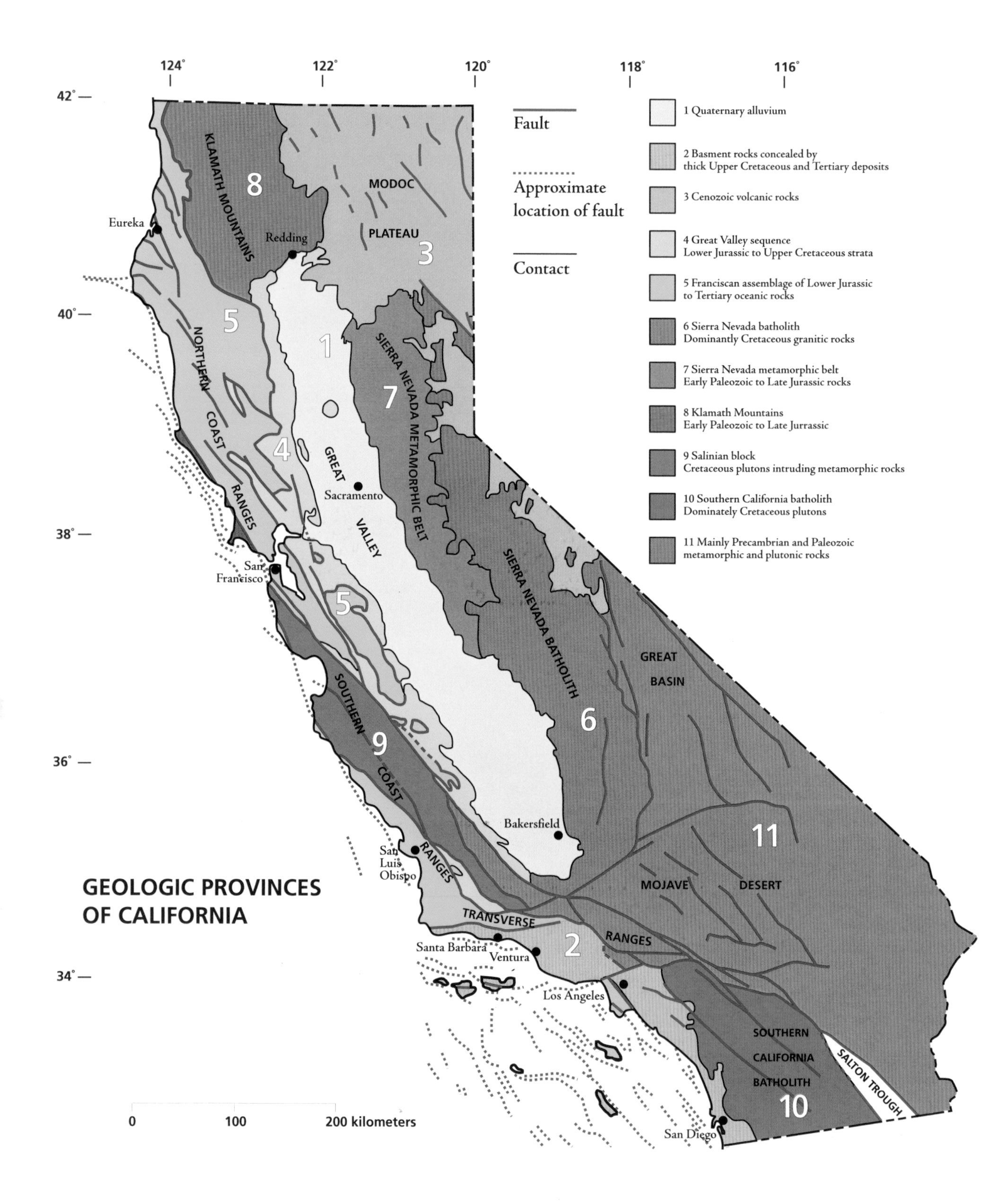
124°
122°
120°
118°
116°
42°
40°
38°
36°
34°
Fault
Approximate location of fault
Contact
1 Quaternary alluvium
2 Basment rocks concealed by thick Upper Cretaceous and Tertiary deposits
3 Cenozoic volcanic rocks
4 Great Valley sequence Lower Jurassic to Upper Cretaceous strata
5 Franciscan assemblage of Lower Jurassic to Tertiary oceanic rocks
6 Sierra Nevada batholith Dominantly Cretaceous granitic rocks
7 Sierra Nevada metamorphic belt Early Paleozoic to Late Jurassic rocks
8 Klamath Mountains Early Paleozoic to Late Jurrassic
9 Salinian block Cretaceous plutons intruding metamorphic rocks
10 Southern California batholith Dominately Cretaceous plutons
11 Mainly Precambrian and Paleozoic metamorphic and plutonic rocks
KLAMATH MOUNTAINS
MODOC PLATEAU
Eureka
Redding
NORTHERN COAST RANGES
SIERRA NEVADA METAMORPHIC BELT
GREAT VALLEY
Sacramento
San Francisco
SIERRA NEVADA BATHOLITH
GREAT BASIN
SOUTHERN COAST RANGES
Bakersfield
San Luis Obispo
MOJAVE DESERT
TRANSVERSE RANGES
Santa Barbara
Ventura
Los Angeles
SOUTHERN CALIFORNIA BATHOLITH
SALTON TROUGH
San Diego
0
100
200 kilometers
GEOLOGIC PROVINCES OF CALIFORNIA

For all of its notoriety near San Francisco, the San Andreas Fault actually must share the seismic limelight with parallel faults both to the southwest (the San Gregorio Fault), and to the northeast across the San Francisco Bay. Averaged over the last couple of hundred thousand years, the southern end of this section of the San Andreas Fault has only moved six-tenths of an inch per year. The other faults take up much of the difference between this rate and the overall two-inch-per-year relative motion of the Pacific and North American plates. The Hayward Fault runs right up the East Bay, cleanly bisecting the University of California at Berkeley's football stadium. David Schwartz, a geologist with the US Geological Survey, has assembled evidence that at least four, and probably seven, major earthquakes have ripped this area in the last twenty-three hundred years. The most recent occurred in the late 1770s. An M=7.0 or greater earthquake would have been bad enough when only a handful of people lived here; with three million now perched directly on top of the Hayward Fault, how much greater a catastrophe will its next large break cause?

The San Andreas in northern California has been quiescent since 1906; meanwhile, strain has been inexorably building within this seismic gap. San Francisco and the peninsula were badly shaken in 1989 by the Loma Prieta, an M=7.1 earthquake originating just north of Santa Cruz beneath the Forest of Nisene Marks State Park. But seismograph records suggest that the Loma Prieta may have actually propagated along a subsidiary fault, not the San Andreas itself. If so, the San Andreas was only minimally relaxed by this earthquake. The clock keeps ticking.

The mountains above Santa Cruz, part of the Coastal Ranges, have an exotic past. Most rocks that lie immediately east of the San Andreas in northern California belong to the Franciscan assemblage, a gaggle of graywacke, shale, chert, limestone, and volcanics. Much of the Franciscan assemblage was first deposited in a deep trough, eventually reaching cumulative thicknesses of 50,000 feet and a total volume of 350,000 cubic miles, enough to cover all of California to a depth of 10,000 feet. Only one type of trough could be this deep: a trench in front of a subduction zone.

The origin and evolution of the Franciscan assemblage is intimately related to plate tectonic processes. Its sedimentary components were deposited in a trench, but other parts were added by processes of accretion. Some volcanic elements of the Franciscan first

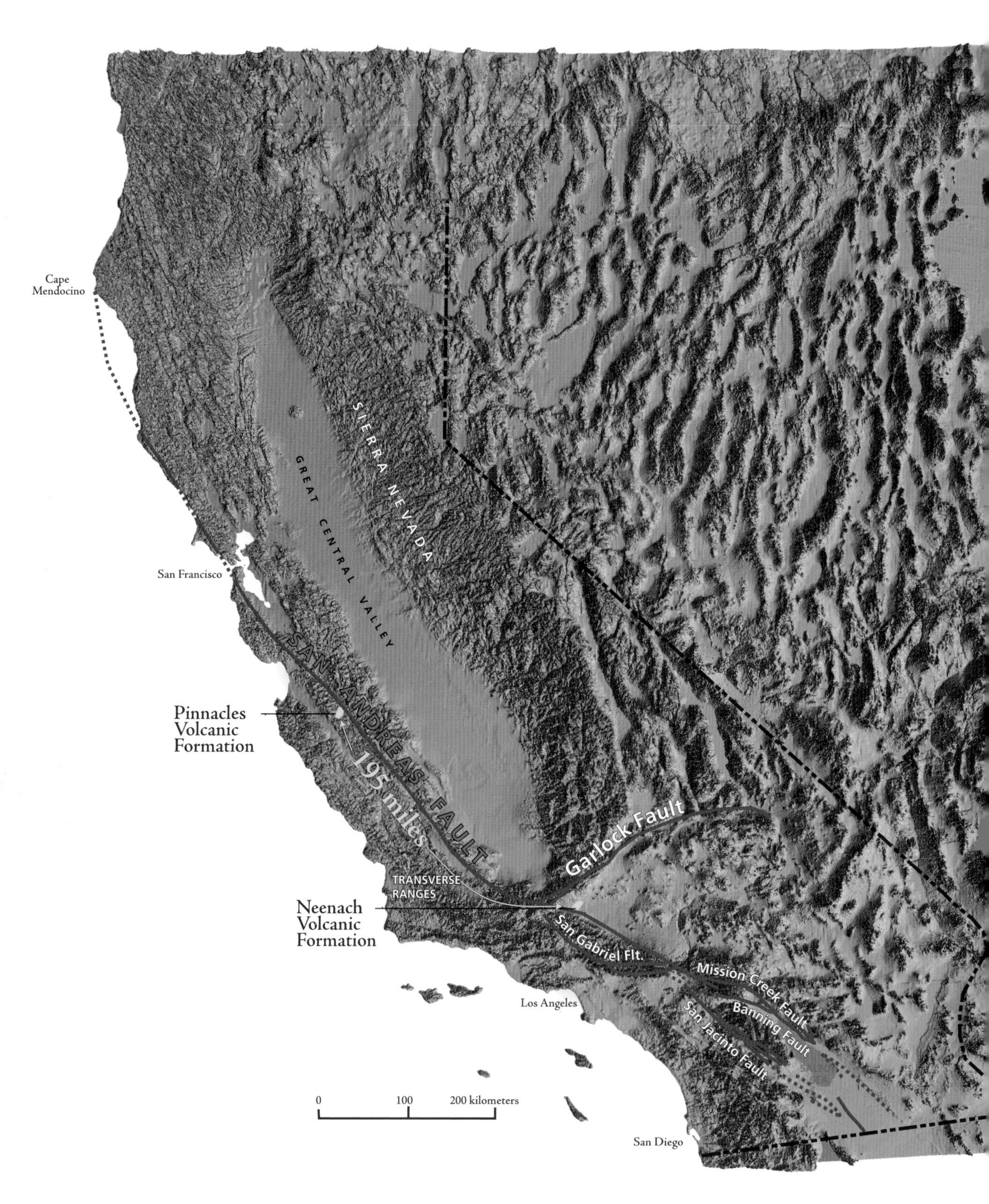

The volcanic rocks of Pinnacles National Monument and those found near the community of Neenach in the western tip of the Mojave Desert were formed 23 million years ago as a single unit. They have since been split and carried 195 miles apart by motion along the San Andreas Fault.

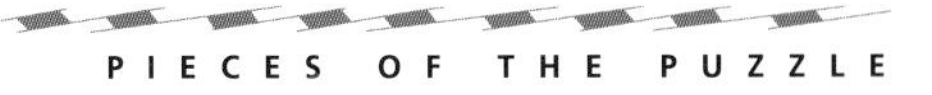

Franciscan cherts, Marin Headlands, Golden Gate National Recreation Area

erupted 1,250 miles to the south of their current position and were transported northward across the eastern Pacific Ocean aboard oceanic crust that has long since disappeared beneath the North American Plate. Sediments shed by mainland erosion were mixed with these incoming pieces of oceanic crust. The Franciscan assemblage was scraped off the top of the subducting oceanic plate and came to rest as a jumbled mess against the western edge of the overriding continent. The worst of the jumble is affectionately known as the Franciscan melange. The Franciscan assemblage is an excellent example of what geologists now call an exotic terrane—rocks that originated elsewhere and now reside alongside other rocks with very different histories.

Central California

Be Here When It Happens! Parkfield, California, The Earthquake Capital of the World!

Personally, I'd be happy to be in Parkfield with or without an earthquake. Empty two-lane roads winding through soft brown hills, full oaks shading deep green grass. California doesn't get more tranquil or more beautiful. The titillating possibility of an earthquake is only value added. Without freeway interchanges or swaying skyscrapers poised to fall, the experience would be more instructive—and far less terrifying.

Parkfield and the San Andreas Fault

Central California is drier than the northern coastline. Here, the evidence of tectonic processes doesn't melt so quickly into the ground. The San Andreas is well exposed in this central, relatively straight section that stretches from San Juan Bautista past the Carrizo Plain. The northern half is characterized by steady, almost unnoticeable, creep, while the southern half is given to episodic slip. Whether by creeping or slipping, the San Andreas here has managed to move material a long way. The volcanic rocks at Pinnacles National Monument have been shipped 195 miles to the northwest since being stripped from their twin, the Neenach Volcanics, at least sixteen million years ago.

Two types of rock generally border the San Andreas in this region. The Great Valley sequence lies to the northeast, and the Salinian Block, to the southwest. The Great Valley sequence began to accumulate in late Jurassic time (about 150 million years ago) when mountains to the north and east shed mud and sand on to what was then the floor of the Pacific Ocean. Much of that seafloor was a package of volcanic rocks, ophiolites, that had previously formed at a mid-ocean spreading center: from bottom to top, these sequences include ultramafic peridotite, gabbro, diabase dikes, and finally, pillow basalt that chilled as it was extruded underwater. Radiometric dating of Coast Range ophiolites yields ages ranging from 165 to 153 million years. Unlike the discombobulated Franciscan assemblage, much of the Great Valley sequence remains coherently stacked up in pretty much the order it was laid down.

The Salinian Block is primarily granite and metamorphic schist and gneiss, distinctly different than the sedimentary Great Valley sequence. Ages typically range between 120 and 70 million years. The most northern exposure of Salinian rocks is at Bodega Head, just beyond Tomales Bay. Salinian granite looks remarkably similar to the granite of the Sierra Nevada, but paleomagnetic evidence raises the possibility that it may have been rafted into California from a point of origin fifteen hundred miles to the south. When first formed, granitic or metamorphic rocks are tremendously hot and will bake any adjacent rocks. The shale and sandstone of the Great Valley sequence show no sign of such contact metamorphism, which implies that the Salinian Block was hauled into position along the San Andreas Fault long after cooling.

The San Andreas Fault has certainly been active in central California, whether by creep or episodic right-lateral slip along the primary fault trace. But the region has been significantly affected by other types of faults. Coalinga was heavily damaged by an M=6.5 quake in 1983, when a steeply dipping buried fault suddenly failed. The epicenter was nine miles northeast of Coalinga, clearly separate from the San Andreas Fault. Its effect was to thrust a wedge of Franciscan melange into and beneath the Great Valley sequence. The fault did not break through to the surface, but a nearby hill called Anticline Ridge was measurably uplifted by the earthquake. This motion necessarily involved a significant degree of regional compression oriented sixty-five degrees oblique to the San Andreas Fault. Such compression, which folds rocks into hills such as Anticline Ridge, occurs simultaneously with other types of motion along the fault.

This satellite image of the Big Bend area, about sixty miles northwest of Los Angeles, shows the San Andreas as well as associated faults, including the Garlock, San Gabriel, San Francisquito, and others; the image was generated using Landsat Thematic Mapper technology.

Big Bend and Transverse Ranges

No sooner had Europeans entered California than they sensed the earth dancing beneath their feet. Camped along the banks of the Santa Ana River on their way through the Los Angeles Basin in 1769, Gaspar de Portola and his soldiers were shaken by a violent temblor "that lasted about half as long as an Ave Maria." They were treated to no less than a dozen aftershocks during the next six days. Welcome to Los Angeles, Don Gaspar; please pass my helmet and armor.

The Los Angeles Basin is immediately south of both the Transverse Ranges and a 180-mile-long segment of the San Andreas that is twisted thirty-five degrees away from the rest of the fault system. Up to this point, it has been appropriate to consider the San Andreas as a reasonably straightforward boundary between two shearing masses—the Pacific and North American plates. But now that we have arrived in southern California, things get a little more complicated.

South of the Carrizo Plain, the San Andreas bends to an orientation of N75°W as it enters the Transverse Ranges. These mountains generally trend east/west, running against the north/south grain of most mountain ranges in California and the rest of the western United States. The Transverse Ranges show evidence of compression and uplift that began within the last few million years and continues today.

As it rolls through this "Big Bend," the San Andreas changes character. To the north, it primarily functions as an agent of intracrustal shear. To the south, right-lateral strike-slip continues across the fault. But because fault orientation rotates by thirty-five degrees in this segment, northwest movement of the Pacific Plate here produces much more compression across the fault. This compression drives the folding and uplift of the Transverse Ranges. The Santa Monica, Santa Susana, and San Gabriel mountains have rocketed skyward over the last two to four million years, driven by the wedging of north-directed thrust faults. Uplift at rates of one-tenth of an inch per year can build mountains in a hurry.

The strongest earthquake to roll through California in recorded history occurred in 1857. Its hypocenter was near Parkfield. Unlike the 1906 San Francisco quake, which ripped simultaneously to the northwest and southeast, the M=7.8 1857 quake extended only to the southeast. Perhaps the fracture would have propagated in both directions had it not immediately encountered the creeping section of the fault that appears to be incapable of building up much strain. Unlike 1906, the 1857 fuse was lit at one end rather than in the middle.

The 1857 fracture raced through the Big Bend, past the sleepy army outpost at Fort Tejon, and on to Wrightwood, a hamlet near Cajon Pass. A small town named Los Angeles, population four thousand, lay thirty-seven miles away from the fault and was jostled but not heavily damaged. Thirty-one feet of slip occurred on the Carrizo Plain. It's hard to imagine thirty-one feet of instantaneous

The Garlock is a left-lateral strike-slip fault that slices across the northern Mojave Desert. Unlike land influenced by the San Andreas, streams here are offset to the left, not right.

slip, but let's try. Stand on a sidewalk facing a row of houses. Suddenly shift not just the houses you see in front of you, but the entire world behind them, ten yards to the right. Let the dust settle and then see what's left of your previously static view of the world. What is remarkable about the 1857 Fort Tejon earthquake is that it barreled right through the Big Bend. Strike-slip movement around the bend should have induced tremendous compaction of rocks on the inside of the bend, and significant extension of rocks on the outside, effects that would limit slip on the fault. But the 1857 earthquake rolled right on through.

The Big Bend also introduces complexities on a larger regional scale by changing the local orientation of the San Andreas relative to the overall motion of the Pacific and North American plates. Thrust faults slice and dice the crust to accommodate compression of the Pacific Plate as it rams into (rather than sliding by) this prow of North America. These faults commonly lie at depths of six to twelve miles, near the interface of the crust and underlying mantle. Slip on such faults may be substantial but frequently does not break through to the surface; thus the name blind thrusts.

Blind thrusts have become a more urgent problem since 1857, primarily because Los Angeles is no longer a tiny village. The suburb of Northridge was ground zero for the 1994 M=6.7 event that ruptured a buried 166-square-mile thrust fault surface in eight seconds flat. Thousands of buildings were badly damaged; sixteen hundred were cordoned off as unsafe to enter. Fifty-seven people died. Tens of thousands of landslides were triggered within the surrounding 400 square miles. Dust from the landslides carried spores of a fungus called *Coccidioidomycosis imitans* and precipitated an epidemic of Valley Fever; three more people died. Damages ultimately reached $40 billion.

As it passes through the Transverse Ranges, the San Andreas is complex, not just geometrically but also temporally. The San Andreas has not always been this region's dominant fault. Up until thirteen million years ago, the three segments of the Clemens Well-Fenner-San Francisquito Fault accommodated a total of sixty miles of right-lateral strike-slip movement. The San Gabriel Fault subsequently slipped at least another twenty-six miles. These older faults are now inactive, frozen since the San Andreas Fault assumed the lion's share of southern California's transform motion five million years ago. Today, strike-slip motion occurs not only on the San

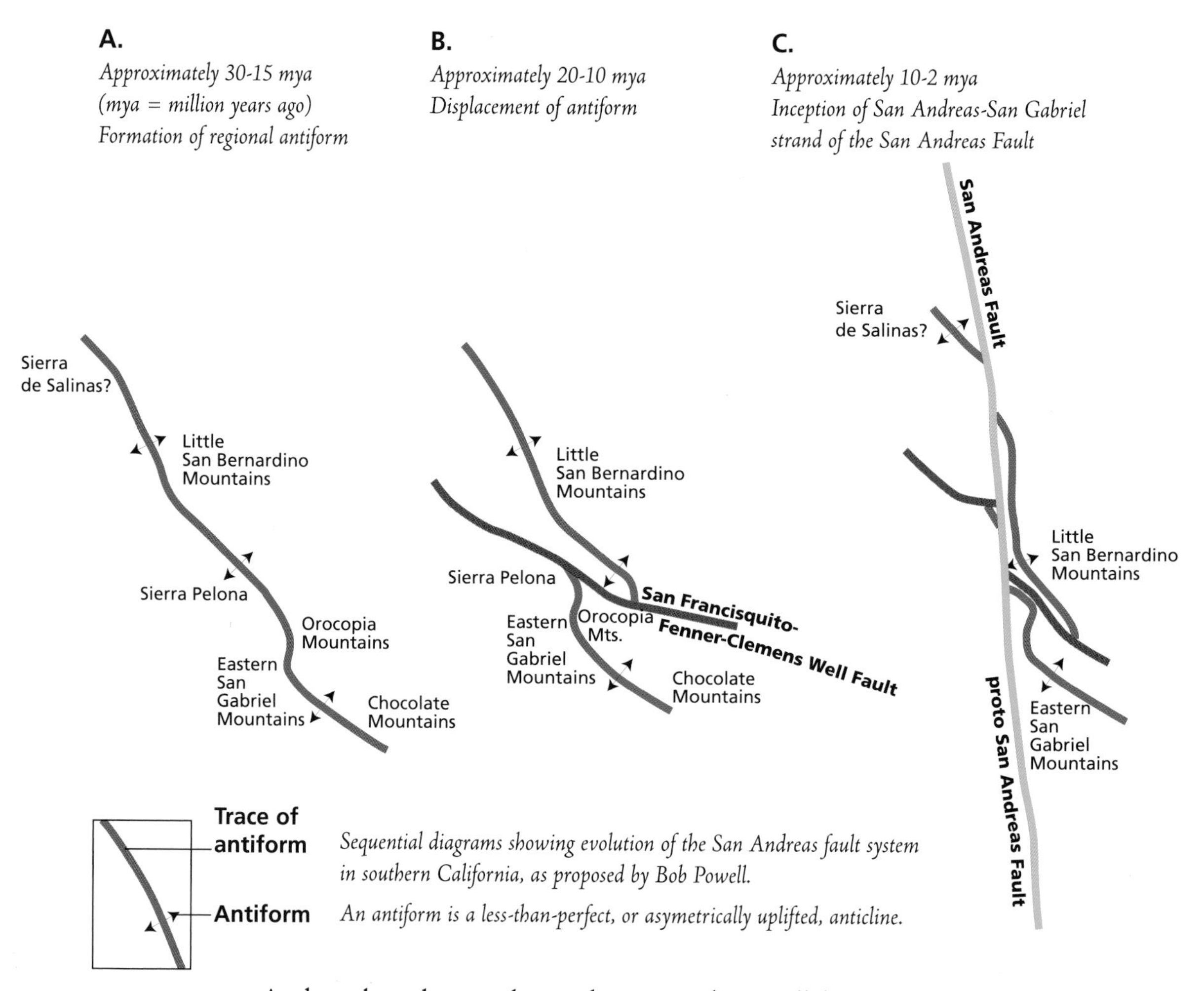

Sequential diagrams showing evolution of the San Andreas fault system in southern California, as proposed by Bob Powell.

An antiform is a less-than-perfect, or asymetrically uplifted, anticline.

Andreas but also to a lesser degree on the parallel San Jacinto and Elsinore faults.

Rocks in southern California are at least as confusing as anywhere else in the state. The predominant petrographic theme is similar to that of the Sierra Nevada: Mesozoic granite intruding Precambrian metamorphic and plutonic host rocks. Mesozoic Pelona schist has received attention from a lot of geologists because faulting has dabbed it across widely separated areas of southern California. Good exposures of Pelona schist are found on the south side of Cajon Pass. On the north side of the pass, sediments have been dumped from the San Gabriel Mountains, across the San Andreas Fault, and onto the broad pediment down which Interstate 15 rolls onto the Mojave Desert. Major sedimentary basins formed throughout southern California as the San Andreas Fault evolved—the Los Angeles Basin is probably the most well known. Farther west, the Ventura Basin accumulated fifty-eight thousand feet of sediment as its bottom warped downward over the last

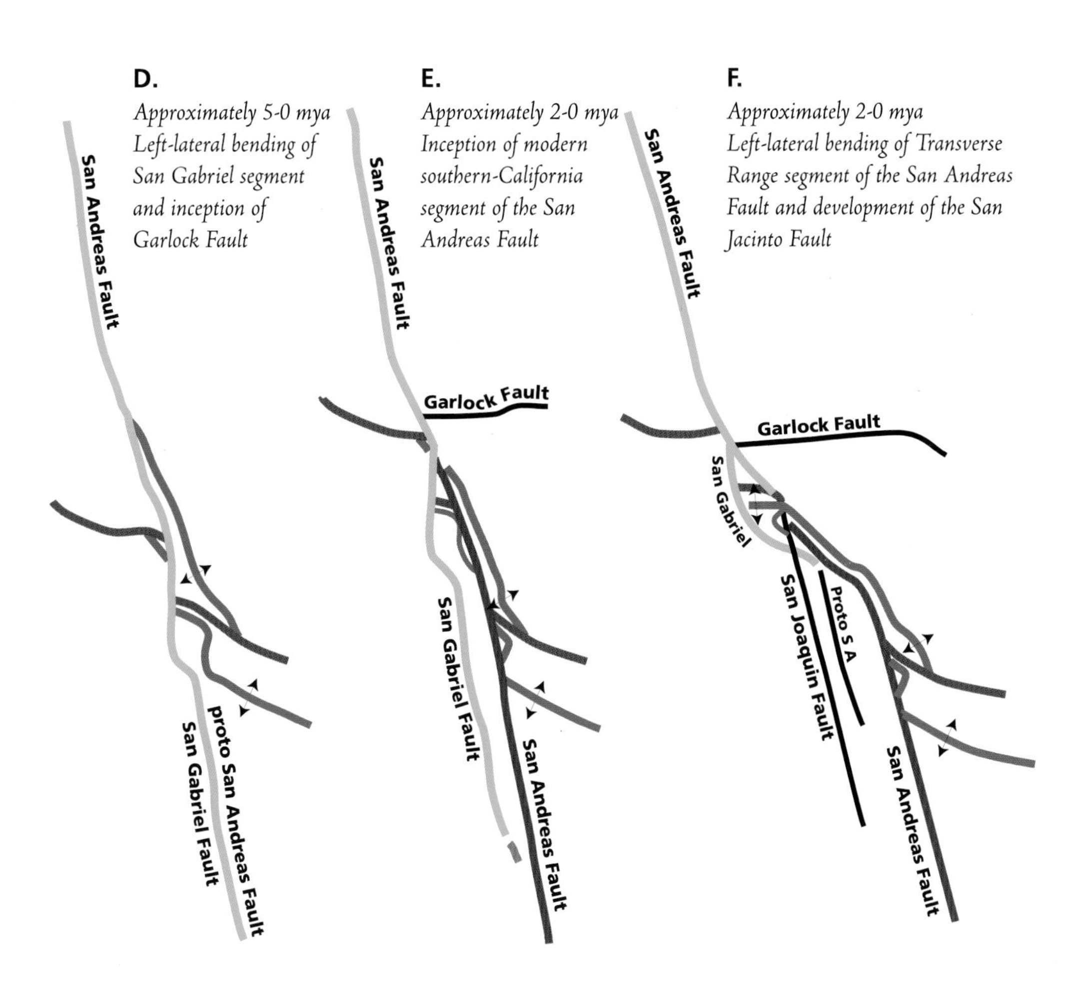

seventy million years in response first to Cretaceous and early Tertiary subduction, then the Miocene extension, and now (the last five million years), compression, folding, and uplift.

In his 1908 review of the San Andreas, Andrew Lawson showed the fault ending just east of the San Bernardino Mountains. Life was simpler in those days. Modern maps now extend it to the Salton Sea. No wonder Lawson was confused. Abeam the town of San Bernardino, the San Andreas looks more like a shattered windshield than a single coherent fault system. It splays into a myriad of parallel and not-so-parallel sub-faults: the Arrowhead, Mill Creek, Mission Creek, Crafton Hills, Santa Ana, Banning, San Gorgonio Pass, Pinto Mountain, Morongo Valley, Garnet Hill, and Vincent. Past Palm Springs, it regroups as the north and south branches of the San Andreas, and at Indio settles down to just being the good old San Andreas as it cruises through the Coachella and Imperial valleys on its way to the Salton Sea.

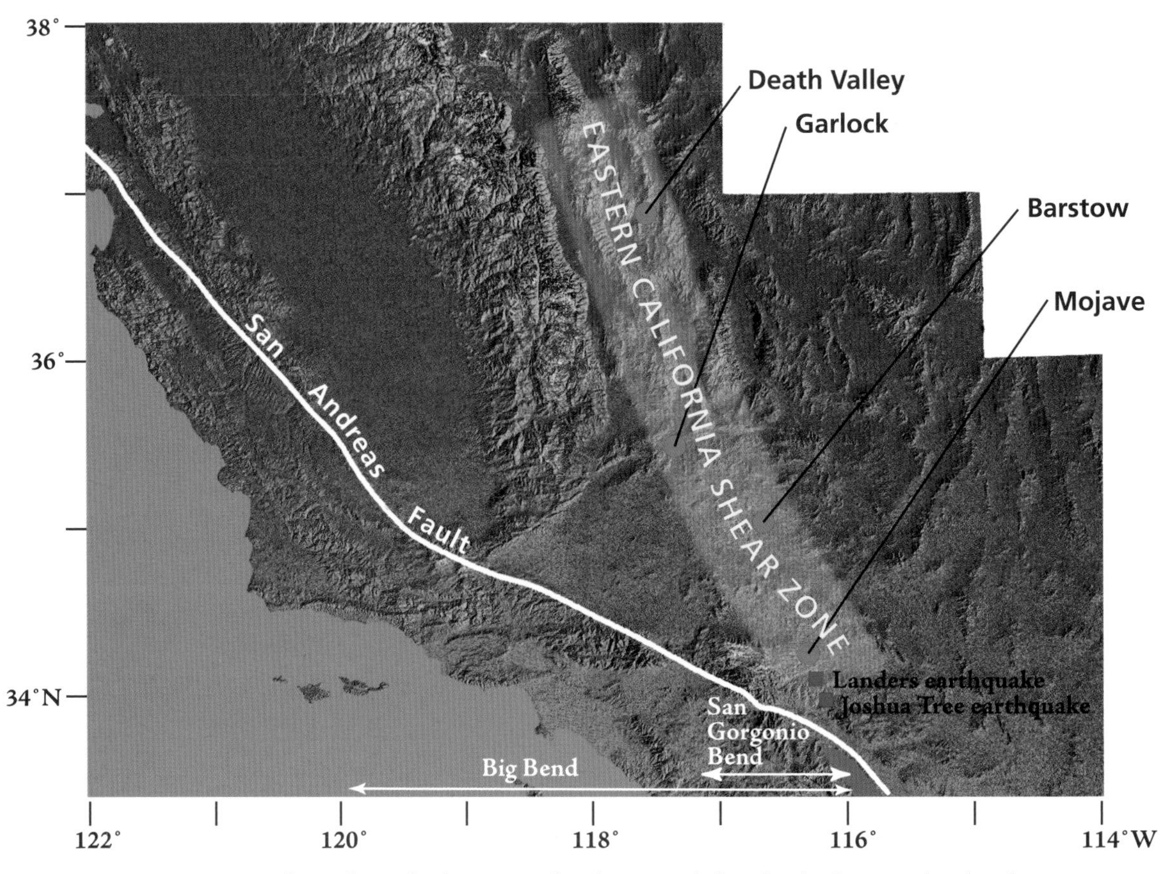

The Landers and Joshua Tree earthquakes were right-lateral strike-slip events that aligned with other right-lateral faults stretching north across the Mojave Desert and through Death Valley along the Eastern California Shear Zone. There is speculation that the principal boundary of right-lateral motion between the Pacific and North American plates will one day abandon the San Andreas Fault and concentrate here instead.

The Mojave Desert and Salton Sea

The Mojave Desert is a triangular wedge which, when seen on a large-scale topographic map, seems to be effectively prying the Sierra Nevada away from the rest of southern California. The Mojave is a distinct physiographic province, typically three thousand feet above sea level and excruciatingly arid. Its northern boundary is the Garlock Fault, arguably California's second-most-prominent strike-slip fault. In a land dominated by right-lateral strike-slip faults, the Garlock is one of only a handful that have left-lateral movement. The Garlock Fault intercepts the San Andreas very close to the enigmatic Big Bend. The Mojave is sliced by other left-lateral strike-slip faults such as the Pinto Mountain and Blue Cut faults of Joshua Tree National Park. Bill Dickinson has suggested that these left-lateral transform motions have connected with those of the right-lateral San Andreas and the southern California thrust faults to rotate great blocks of the Transverse Ranges and Mojave Desert, all pinwheeling in clockwise circles.

The 1992 M=7.6 Landers earthquake was the greatest shock to roll through California in forty years. Relatively few people reside close to its epicenter north of Joshua Tree, and thus, damages were under $100 million. There has been such a concentration of earthquake activity in this part of the Mojave Desert, however, that some geophysicists are beginning to wonder if we are witnessing early signs of a jump of plate motion from the San Andreas to a line that stretches through Joshua Tree, Death Valley, and north toward Mount Shasta.

At the beginning of this century, there was a Salton Sink but no Salton Sea. The Sink was three hundred feet below sea level, bone-dry except when it rained. And it rarely rained. To would-be farmers, this looked too good to be true. They figured that all they had to do was get a little bit of the nearby Colorado River headed toward the sink, plant seeds, and then jump back before the broccoli and lettuce ran them over. The California Development Company tapped a canal into the Colorado River just north of the international boundary in 1901. It took a while to get the bugs out: dredging the silt that perennially clogged the canal, tinkering with the headgates,

The Salton Sea

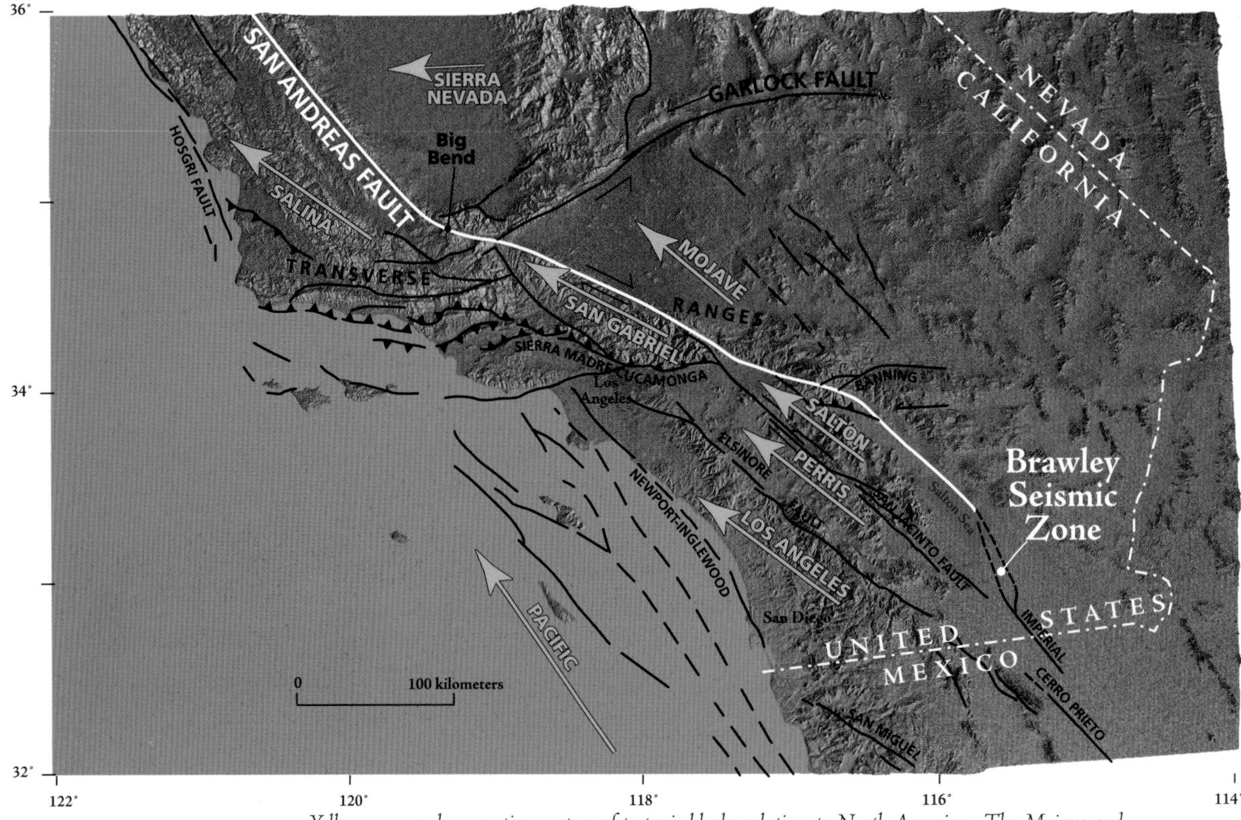

Yellow arrows show motion vectors of tectonic blocks relative to North America. The Mojave and Sierra Nevada blocks lie to the east of the San Andreas Fault and thus are part of North America, but Basin and Range extension and Mojave rotation moves these blocks to the northwest.

and luring a few more farmers. But in 1905, the company finally got it right. Too right. Knowing a good thing when it saw one, the entire Colorado River roared into the canal, washing out all the headgate fixtures and with them, any hope of control. The Salton Sink became the Salton Sea. Two years later, the river was finally wrestled back into its old channel leading to the Sea of Cortez.

Should we have expected anything else? Not really. The Colorado River has been flipping back and forth across this delta like an out-of-control fire hose for the past two or three million years. The Coachella Valley and Salton Sea are parts of an extensional basin whose floor is actively stretching and subsiding. Drillers can sink bore holes two and a half miles into sediment deposited by the Colorado River without reaching basement rock. Geophysicists have identified two seismic reflective surfaces, the first at a depth of three miles and another at seven-and-a-half miles. The rocks between the two reflectors are most likely the metamorphic products of heating and compression of the basin's sediments. The lower surface probably represents true basement rock.

The Colorado River has flushed sediment into this basin many, many times in the past. Indeed, a good argument has been made that the Colorado, at some point unable to find an outlet to the Sea of Cortez (perhaps because it didn't exist), once made its way to the Pacific Ocean by going west through Banning Pass and emptying into the Los Angeles Basin. For now, the Colorado River is behaving itself, bypassing Palm Springs and the Imperial Valley. The Salton Sea, replenished only by irrigation run-off, is evaporating and growing saltier every day. Already it is much more saline than sea water. Stock prices for the California Development Company and its successors continue to decline.

Where the San Andreas Fault leaves off, the Brawley Seismic Zone and Cerro Prieto Geothermal Area pick up. These northeast-trending structures appear to be small, on-shore spreading centers, as evidenced by recent volcanic activity replete with geothermal mudpots and an underlying hot, thin crust. Like normal warm-blooded spreading centers, both of these are connected by faults accommodating transform motion: the San Andreas, Imperial, and Cerro Prieto. This transform motion runs down the length of the Sea of Cortez, interrupted now and then by additional seafloor spreading centers, until it reaches the Rivera Triple Junction offshore of Mazatlan. This tectonic arrangement has succeeded in stripping Baja California away from mainland Mexico, and is carrying the peninsula northwest along with the rest of coastal California.

Basin and Range

One last segment of the San Andreas system remains to be examined. Perhaps its inclusion stretches the system's definition, since this Basin and Range Province starts on the eastern side of the Sierra Nevada and continues all the way to central Utah. That's a long way from Petrolia, Hollister, and the Salton Sea. But consider the 1872 M=7.6 Lone Pine earthquake. It shook up most of California and absolutely flattened the Owens Valley town of Independence. John Muir was treated to a spectacular Yosemite Valley rockfall triggered by the quake. Owens Valley continues to be the locus of earthquakes on an almost-weekly basis.

The Basin and Range is a broad zone of the western United States that began to extend in Miocene time, about seventeen million years ago. This extension occurred as blocks of basement rock pulled apart, not unlike a deck of cards being fanned out onto a table. The crust

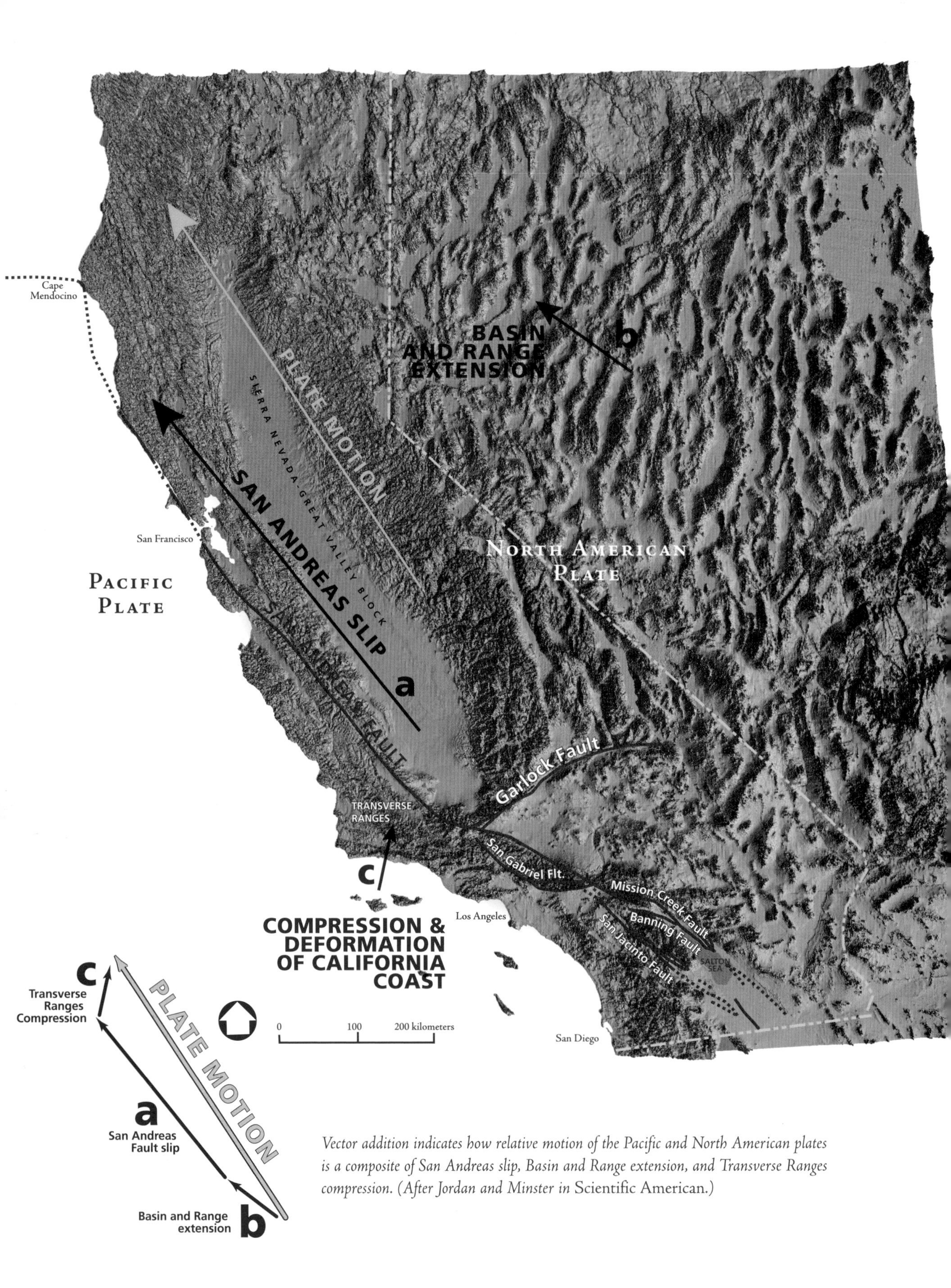

Vector addition indicates how relative motion of the Pacific and North American plates is a composite of San Andreas slip, Basin and Range extension, and Transverse Ranges compression. (After Jordan and Minster in Scientific American.*)*

Cinder cones beneath the eastern Sierra Nevada, north of Independence, California

Normal fault created as Death Valley slides down the face of the Black Mountains

beneath the Basin and Range has been thinned by 30 percent; it now transmits more heat from the underlying mantle than any other province in North America. The cumulative effect of this thinning has been to transport the Sierra Nevada 165 miles northwest, away from the Colorado Plateau on a N75°W trajectory, at a rate of just under half an inch per year.

During the 1970s, the early true believers of plate tectonics theory had been stumped by the discrepancy between observed movement on the San Andreas Fault and computed movement between the Pacific and North American plates. The Pacific Plate is moving 1.9 inches per year, heading in a direction of N36°W relative to North America. But, measured at Parkfield, the San Andreas Fault is moving along a line oriented N41°W at a rate of 1.3 inches per year. How can we correct for this difference between fault and plate motions? Bernard Minster and Thomas Jordan proposed that the discrepancy could be resolved by considering not only the motion of the San Andreas and its subsidiary strike-slip faults, but also integrating the compression of the Transverse Ranges and the northwest extension of the Basin and Range. These three very different tectonic processes have been operating simultaneously since the Pacific Plate began to slide along the western edge of North America. Taken together, they begin to account for the observed plate motions.

In the Field: David Schwartz

DAVID SCHWARTZ CARRIES A BEEPER that goes off any time northern California is tickled by an M=3.5 or greater earthquake. It goes off a lot. His wife has threatened to take a hammer to the beeper. Schwartz is chief of the San Francisco Bay Area's Earthquake Hazards Project. After each major earthquake, he or another member of the project goes the site to inspect damage and assess possible evolving hazards. Is a landslide about to shake loose into a man-made lake and create a wave that would overtop its dam? What is the likelihood of an aftershock delivering a crippling blow to an already-weakened pipeline?

Schwartz joined the US Geological Survey in 1985 after spending a dozen years doing private-consulting geology. His great love is the study of faults and earthquakes. We met in Berkeley, intending to visit sag ponds along the Hayward Fault. We drove along five-lane freeways, past thousands of businesses, and through dense clusters of houses from El Cerrito to Fremont. "In northern California," he observed, "the only certainties are death, taxes, and earthquakes." When (not if) the Hayward Fault slips again, these buildings will be violently shaken; many will fall. Schwartz was shocked to see how closely condominiums had encroached upon Tule Pond in Fremont. It is a sag pond, clearly delineating the trace of the Hayward Fault. Standing alongside the fault, I could have easily hit homes on either side with a rock.

Day to day, urban dwellers live in the sobering shadow cast by the San Andreas Fault system. But urban sprawl is obliterating surface evidence of the few fault traces that remain available for geologists to study. In 1997, Schwartz looked around for undisturbed land along the fault. He found it in a low spot along the second fairway of El Cerrito's Mira Vista Golf Course, which was built back in 1912. He was granted ten days to trench two gashes into a sag pond, a project that would have normally taken months. Digging down as much as ten feet, he found evidence of at least four, and possibly as many as seven, major earthquakes. Carbon-14 dates run by the Lawrence Livermore Laboratories confined these events to the last twenty-three hundred years.

Schwartz tries to reconstruct earthquake recurrence intervals to gain insights into fault mechanics. He has worked in trenches up and down the San Andreas; he has traveled to Mongolia to study comparable faults and earthquakes of the Gobi Alti Desert. Schwartz feels that the Survey offers a good vantage point from which to ponder the larger geologic questions whose answers may take years to piece together.

David Schwartz

I asked Schwartz if he knew of any accounts of people actually observing the ground rupture during an earthquake. He hadn't heard of any in California, but did recall the story of two hunters who saw the ground shred during an Idaho earthquake. He was quiet for a second before confiding that he has awakened smiling on more than one occasion from dreams of an earthquake that silently unzips the ground into a fresh white scarp. "You know, I would hate to see the destruction and suffering caused by a major earthquake in the East Bay—but I would give anything to be there when it happened." We were standing right on a scarp of the Hayward. I took a few steps backward.

Earthquakes: Prediction or Planning?

"So when is the next big one gonna hit, doc?" Few questions can simultaneously so tantalize and frustrate a geologist. Physicians trying to navigate grocery-store aisles adroitly deflect Mrs. McGillicutty's eternal questions about her lumbago. Party-going stock brokers cringe each time another reveler tries to elicit a hot new investment tip for free. But geologists—who tend to be straightforward people with a preference for plaid flannel shirts—just stand there and stammer.

"The next big one? Well, there's a 39 percent chance of at least a magnitude 7.2 event occurring within fifty-three kilometers of here in the next seventeen years. Though we don't know for sure, and it could be any time." Let's face it: no one is satisfied with this answer, but so far, playing the odds is the best we can do.

Pacaimo Reservoir perched above the San Fernando Valley near the San Gabriel Fault

What we do know is that earthquakes tend to cluster in time. Foreshocks and aftershocks do bracket major quakes, but geologists are only able to classify quakes as one or the other after the entire sequence is over. If an M=6 event occurs in southern California, history has shown that there is a 6 percent chance that it will turn out to be the foreshock to an even larger earthquake within the next six days. Aftershock behavior tends to be more reliable than that of foreshocks. Omari's Law—named for its Japanese promulgator—states that the number of aftershocks will decrease in inverse proportion to the amount of time since the main shock. This makes intuitive sense: the earth is settling in after being disturbed, tossing and turning under the covers before going back to sleep.

Earthquake prediction has certainly been geology's Holy Grail for many years. Yes, it would be nice to discover a new vein of molybdenum here or name a new rock formation there. But earthquake prediction is in a different league, as accurate forecasts could save hundreds of thousands of lives. Long-range predictions (a.k.a. "probabilities") have been fairly successful, but reliable short-term predictions have so far proven elusive. All sorts of angles have been tried; some even have a conceivable basis in reality. Geysers bubbling above a fault could get more active as rocks underneath begin to shift. Dogs could bark louder or stray farther, anticipating something that we can't sense. The atmosphere could became calmer before an earthquake strikes. But they don't—at least not in the identifiable patterns necessary to make meaningful predictions.

For an earthquake prediction to be useful, it must be geographically limited. It does little good to say that a large earthquake might occur somewhere in California. If a quake predicted for the entire state severely shakes only Coalinga (population eighty-two hundred), the folks of Los Angeles and San Francisco might not respond to the next warning with much urgency. To be useful, a prediction must also be timed with fair precision. It must be made in time to allow people to leave buildings, but its value diminishes when it is made too soon. How many hours—or days, or weeks—are people willing to camp out in their yards before moving back inside, back into harm's way?

At least one major earthquake has been predicted with remarkable precision. During the first days of February 1975, a swarm of shocks was felt near Haicheng, east of Beijing, China. By February 4, sixty quakes were rattling the seismographs every hour.

The Hayward Fault runs beneath Fremont and the East Bay area.

People noticed odd animal behavior, funny-tasting well water, and anomalous ground fog. That afternoon when the foreshocks suddenly and ominously quieted down, Chinese authorities issued an official warning of an imminent major earthquake, and three million people dutifully moved out of their unreinforced buildings and into tents. Hours later, an M=7.3 quake destroyed 90 percent of the city's homes. Sadly, a few hundred people died. Had the warning not been issued, a hundred thousand more would likely have lost their lives.

Seventeen thousand years ago, the San Bernardino Mountains spewed out the Blackhawk Slide, quite possibly triggered by an earthquake along the San Andreas. Imagine the result if this slide—four miles long and a mile-and-a-half wide—were to collapse south into the Los Angeles Basin today.

This single great success was overshadowed by a greater tragedy just one year later. The M=8.0 T'ang-shan earthquake struck ninety-five miles southeast of Beijing with no premonitory foreshocks. General warnings about earthquake potential had been issued for the area over the preceding year or two, but no specific location or date was given. This time, at least a quarter-million people died.

So for the time being, we are left with just long-term earthquake prediction, primarily based on the examination of patterns of previous seismicity. These patterns can be deduced from historical records as well as prehistoric evidence of repeated fault movement. Someday, GPS-based plate motion studies might be used to identify areas of the crust that are bending but have not yet broken. For now, we're given long-term statistical probabilities with wide margins of error. For instance, if you live in California, it's a safe bet that you will feel an earthquake sooner or later. A little more specifically, if you live in the San Francisco Bay Area, geologists at the USGS are giving two-to-one odds that you will experience an M=7.0 event in the next thirty years, originating on one of the three most active faults in the area: the San Andreas, Hayward, or Rogers Creek. In Los Angeles, it's even money on a similar-sized shock in the next five years.

As earlier described, in 1985, the USGS put its money where its mouth was by predicting an M=6.0 or greater earthquake in the vicinity of Parkfield, California, some time before 1993. This forecast was based on observed twenty-two-year cycles between significant earthquakes in an area that had last been shaken in 1966. The Survey and the California Division of Mines and Geology saturated Parkfield with a vast array of seismographs, strong-motion seismometers, tiltmeters, creepmeters, and other equipment designed to "capture" the upcoming quake. *National Geographic* magazine installed cameras in residents' homes to record their reaction to the earthquake. Small quakes rolled through, such as the M=4.7 event in October 1992 that could have been a foreshock to the predicted larger quake. These sent the media into paroxysms of giddy anticipation. But so far, Parkfield is as peaceful as ever. Be patient—down below, the plates are still moving. The longer we wait, the bigger It will be.

How can these predictions be improved? Paleoseismologists are digging more trenches across segments of the San Andreas and its subsidiary faults, and are thus able to construct more accurate estimates of recurrence intervals. In the last few years, geologists have begun to think about the effects of one earthquake on the probability of another earthquake in the future. The mechanics are fairly straightforward: in general, a large earthquake will relieve stress that has built up over time across its fault. But more stress relief will occur in some directions out from the epicenter than in others. In fact, there will be quadrants where local stresses will *increase* after an earthquake.

Ruth Harris and Robert Simpson of the USGS in Menlo Park retrospectively studied the great 1857 Fort Tejon earthquake and wondered if subsequent major earthquakes on or near the San Andreas were encouraged by stress changes induced by the 1857

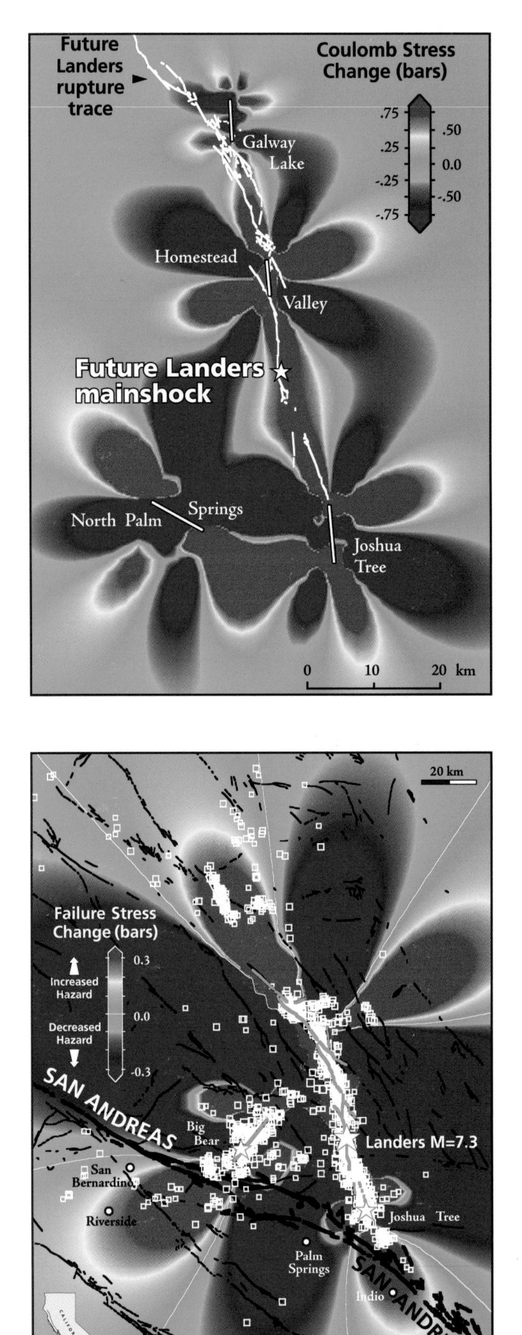

Failure stress changes caused by the four 5.2 shocks within 50 km of the Landers earthquake occurring during the 17 years before the right-lateral Landers rupture. The Landers surface rupture tends to lie within the zone of elevated stress change and is favorably oriented for right-lateral failure (bottom illustration, black lines).

event. What they discovered was that from 1857 to 1907, thirteen out of thirteen M=5.5 or greater quakes observed in southern California were probably hastened by the Fort Tejon quake. They also found that, after fifty years, the influence of the Fort Tejon quake began to fade. When the M=7.6 Landers earthquake struck in 1992, it triggered an M=6.7 aftershock at Big Bear three hours later. The Big Bear Fault was optimally aligned and precisely located within one of the lobes of increased stress generated by the Landers quake. A segment of the San Andreas Fault itself was also within a lobe of increased stress, and over the next few months, experienced an acceleration of small slips due to the Landers event.

These calculations of stress changes are still in their infancy, however, and though they might refine the art of earthquake prediction in both space and time, they do not fundamentally change its nature: for now, prediction remains an issue of odds. Might there be, however, another way to skin this cat? If you can't *predict* a natural event, can you perhaps *cause* it instead? In 1966, geologists in Colorado noticed an unexpected swarm of earthquakes beneath the Rocky Flats Arsenal near Denver; these small quakes correlated exactly with the military's high-pressure pumping of liquid waste deep underground at the arsenal. A light bulb clicked on. USGS geologists set up shop at an all-but-abandoned oil field near Rangely on the western side of

Colorado. Very few people lived nearby. They pumped fluids under high pressure underground, and for the first time in history, intentionally triggered earthquakes.

The implications were staggering. If geologists could set off earthquakes, and if the magnitude of these induced earthquakes could be controlled, then stresses along a fault could be lowered before a catastrophic natural earthquake occurred. For better or worse, this idea has been quietly shelved because there is too great a chance that it might not be possible to control an earthquake once it started. Understandably, the USGS is reluctant to pay to rebuild the entire San Francisco Bay Area or the Los Angeles Basin if things get out of hand.

So while we wait for the art of earthquake prediction to become a more reliable science, what do we do? Plenty. If earthquakes are inevitable, then we might as well plan for them. Michael Rymer notes that we don't throw away seat belts just because we can't predict car crashes. Automobiles have seat belts not so that we can arrange highway accidents ahead of time, but so that drivers have a better chance of surviving accidents. Buildings, like belts, should be designed to withstand a predictable degree of violence.

People are injured and killed not by the ground's shaking during an earthquake but by structures that collapse because of the shaking. During an earthquake, quaint old homes can suddenly become Salvador Dali paintings as door frames twist and ceilings buckle. A strong-motion sensor in Tarzana recorded horizontal ground accelerations that were almost twice that of gravity during the M=6.7 1994 Northridge earthquake. This sort of violent motion, which can instantly turn an ordinarily sedate refrigerator into a deadly missile, suggests that heavy household objects be secured in place. Homes and businesses can be designed—or in many cases, retrofitted—to withstand significant ground movement. Many new large buildings are now constructed with energy-absorbing systems between the building and its foundation; these systems are designed to let vibrations simply roll on through without inducing much damage.

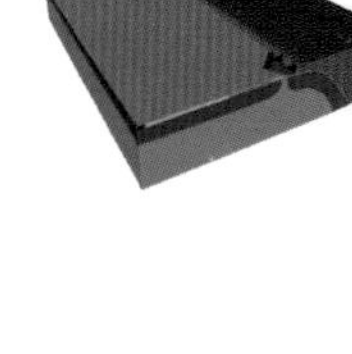

We know that shaking will always be bad on unstable bedrock, worse on unconsolidated soils, and worst of all on mud and fill. During the M=7.1 1989 Loma Prieta earthquake, the Marina district of San Francisco was rocked three to four times as hard as surrounding bedrock areas because it sits on a hundred feet of mud and sand that were dumped into San Francisco Bay decades before.

When shaken, not stirred, sand flows as a liquid and can no longer support buildings situated on it. A three-quarter-mile-long double-decked segment of Interstate 880 south of Oakland also collapsed during the Loma Prieta earthquake, with the loss of forty-one lives. The collapse happened only where the freeway was built on fill. San Francisco subsequently took the hint and razed the damaged multilevel Embarcadero segment of Interstate 280—it too had been built on fill.

Landslides are another common problem throughout California and are often triggered by earthquakes. Their impact can be mitigated by keeping groundwater drained from threatening hillsides before an earthquake hits. Slopes can be stabilized and unstable soils can be made more earthquake-resistant—for a price. More obviously, zoning could prohibit construction on these sites in the first place.

SO WHAT DO WE MAKE of all this? In a piece that appeared in the *San Jose Mercury News*, Paul Reasenberg, a seismologist with the USGS in Menlo Park, reflected on our reaction to earthquakes soon after the 1994 Northridge event:

> *I remember seeing my neighbors after the Loma Prieta (1989) earthquake. Our homes were damaged only very slightly by the earthquake, although some chimneys did fall in our neighborhood. We all talked together out on the sidewalk, comparing notes on what fell down and about how we got home through the traffic chaos, how we lit candles during the power outage and how our kids and pets coped. There was a sense of community. We felt like survivors, and that felt good.*
>
> *I fear that those of us who did experience the Loma Prieta earthquake actually may have developed a false sense of security from it. "We had our earthquake, and we got through it, so what's the big deal?" But we didn't have our earthquake. Santa Cruz had theirs. It was sixty miles away from San Francisco, and while it released five time more energy than last week's L. A. quake, it was centered up in the Santa Cruz mountains, away from most people, buildings and bridges.*
>
> *It's the distance that saved most of us in the Bay area, folks. Unfortunately, I'm afraid that for many of us, the experience ("We got through it, no sweat!") is speaking louder than the seismological warnings. And that's too bad. Because really big, bad earthquakes in densely populated areas don't happen often enough for everybody to learn by experience.*

Modern landslides along the Lost Coast of northern California

Fremont is much more densely developed now than when the Hayward Fault last ruptured a century and a half ago.

We learn to take appropriate precautions. We weigh the chances of an M=7.7 earthquake striking again, engineer our bridges and retrofit our homes, and hope that it doesn't turn out to be an M=8.0 event the next time around. We learn to go on living our lives in the face of what must one day inevitably occur. Being prepared helps us to breathe a little more easily along the way.

The surface of the earth is sculpted by events like landslides, floods, and fires that occur with frequencies that resonate with the rhythms of our lives. Flood scars last for decades, climate patterns, for centuries. Such events change the face of our world every few decades—often enough that each generation adds its experience to a collective human understanding of the processes involved. We use this information to make decisions about how and where we want to live: Seattle because it's green (if you can stand the rain), Tucson because it's dry (if you like the heat).

But life and, so far, earthquakes are unpredictable. Why are we always so surprised when the inevitable happens? Probably because earthquakes catastrophically affect our lives less frequently than other natural phenomena. Truly great shocks can be separated by centuries, and we forget their vehemence. We watch a gaggle of reporters on the six o'clock news flock to a mudslide in Big Sur or a broken levee on the Sacramento River, and listen to them talk in sonorous voices about the Power Of Nature. But when a couple of hundred miles of the San Andreas slips again—and slip it must—we will not be hearing reports of a home flooded here and a road washed away there. Instead, we will be seeing images (if the reporters are still on the air) of destruction on a vastly different scale: hundreds of thousands of buildings will be shaken, millions of lives will be changed.

I met an elegantly dressed elderly gentleman in Thermal, California, one Sunday morning while breakfasting at a restaurant that offered mariscos and fish burritos. When I asked, he told me his name in lovely Castillian Spanish, with seven surnames that tracked his maternal and then paternal heritage back through five generations. Had he lived here long? "Yes, long enough to have felt the earthquake at Imperial in 1940," he said. We talked about earthquakes for a while; I was impressed with the sophistication of his knowledge of fault mechanics. But more importantly, he displayed a wisdom that can only grow from experience. "You know, there is no fear as great as that which we feel when the earth trembles. Strong men cower and the animals all become agitated. I saw that in 1940 and have seen it many times again since then."

TWENTY YEARS AGO when I was studying structural geology at Stanford, I'd have good days and bad days. Sometimes it was all I could do to get on my bicycle and pedal up into the foothills. I would stretch out perpendicularly across what I guessed was the San Andreas Fault, close my eyes, and strain to sense any incipient movement. Sometimes I wiggled just a little bit to see if I could set anything off. The fault didn't budge. In this case, no vibrations were good vibrations. I lay there imagining what the world was like, five and ten miles below. Ever since, I've been trying to see down into the earth, to feel its pulse, to reach into that Third World.

Acknowledgments

MY THANKS to the many people within the US Geological Survey who helped pull this book together: Michael Rymer, Carol Prentice, David Schwartz, and Bob Powell spent time in the field answering my questions and sharing their work. Ross Stein tried to point me in the right direction early on in my research. Jenny Prennace was always helpful as I studied at her USGS library in Flagstaff. Paul Segall at Stanford University shared his ideas about the role of GPS in understanding plate movement. Howard Shifflett and his student Mike Gray enthusiastically showed their study areas near the Salton Sea. David and Suzanne Brown let me explore the faulted land around their lovely Plantation House at Cazadero. Robert Sickler was kind enough to lend me a hard hat while we examined his trench across the Maacama Fault at Ukiah.

Bill Dickinson and William Guyton reviewed the manuscript and suggested valuable improvements. Carole Thickstun and Lawrence Ormsby were wonderful to work with as designer and illustrator. Charlie Money and his sidekick Susan Tasaki, both of the Golden Gate National Parks Association, were supportive at each bend in the road. Bonnie Murchey with the USGS in Menlo Park and Joe Zarki at Joshua Tree National Park helped coordinate this project from the beginning.

I would like to reach back twenty years and also thank the two people who most effectively shaped my view of geology—Charles Barnes at Northern Arizona University, and Arvid Johnson at Stanford University. And finally, as always, Rose Houk's humor, patience, and grace are behind every word I write.

Suggested Readings

Davidson, J., Reed, W., and Davis, P., 1997; *Exploring Earth*: Prentice Hall, 477 pp.

Dickinson, W.R. and Wernicke, B.P., 1997; Reconciliation of San Andreas slip discrepancy by a combination of interior Basin and Range extension and transrotation near the coast: *Geology*, vol 25, no 7, pp 663-665.

Earthquakes and Volcanoes: a serial publication by the USGS for lay readers, with specific issues on major earthquakes as they occur.

http://quake.wr.usgs.gov: a web-site maintained by the USGS for earthquake-related information for the public.

Iacopi, Robert, 1996; *Earthquake Country*: Fisher Books, 146 pp.

Kious, W.J. and Tilling, R.I., 1997; *This Dynamic Earth*: U.S. Geological Survey, 77 pp.

Lawson, Andrew, 1908; *California Earthquake of April 18, 1906*: Carnegie Instituion of Washington, 643 pp.

Moores, Eldridge, 1990; *Shaping the Earth: Readings from* Scientific American, 206 pp.

Sieh, K.E. and Jahns, R.H., 1984; Holocene activity of the San Andreas fault at Wallace Creek, California: *GSA Bulletin*, vol 95, pp 883-896.

Simpson, Robert, ed., 1994; The Loma Prieta, California Earthquake of October 17, 1989: *USGS Professional Paper* 1550-F, 131 pp.

Stein, R.S. and Yeats, R.S.; Hidden Earthquakes: *Scientific American*, June 1989, pp 48-57.

Vogel, Shawna, 1996; *Naked Earth*: Plume Books, 217 pp.

Wallace, Robert, ed., 1990; The San Andreas Fault System, California: *USGS Professional Paper* 1515, 283 pp.

Index

Italicized page numbers refer to photographs or illustrations